DIGG

THE LEG
THE AUSTRAL

DIGGER
THE LEGEND OF
THE AUSTRALIAN SOLDIER

JOHN LAFFIN

Author's Note

I wrote the first edition of this book in 1959 and many historians have quoted from it. This version is not just a reprinting or even a revised edition; it is a complete rewriting. I rewrote the book because my own knowledge of the Australian soldier and of war generally has increased by the experience and research of nearly thirty more years. In addition, the Australian soldier has himself had so many more years of experience, including service in Indonesia and Vietnam.

Copyright © John Laffin 1959, 1960, 1986, 1990

This paperback edition published 1990 by Sun Books
THE MACMILLAN COMPANY OF AUSTRALIA PTY LTD
107 Moray Street, South Melbourne 3205
6 Clarke Street, Crows Nest 2065

Associated companies and representatives throughout the world

This revised edition first published 1986 by
The Macmillan Company of Australia

National Library of Australia
cataloguing in publication data

Laffin, John, 1922–
    Digger, the legend of the Australian soldier.

    Rev. ed.
    Bibliography.
    Includes index.
    ISBN 0 7251 0594 1.

    1. Australia. Army – History. 2. Soldiers – Australia –
    History. 3. Australia – History, Military. I. Title.

355.3'1'0994

Set in Bembo by
Computype Export, New Zealand
Printed in Hong Kong

Contents

Acknowledgements

In writing this history of the Australian soldier I have referred to many military histories; Australian, English and German regimental diaries; and other books. I have studied army files, soldiers' letters and newspaper reports. It is not possible for any historian to write about the Great War (World War I) without using C. E. W. Bean's various volumes of the *Official History of Australia in the War of 1914–1918*; he is the primary source and the ultimate authority. Also invaluable is H. G. Gullett's contribution to the official history, Volume VII, *Sinai and Palestine*. The official histories of World War II, *Australia in the War of 1939–45*, are basic sources: *To Benghazi* (1952) and *Greece, Crete and Syria* (1953) by Gavin Long; *Tobruk and El Alamein* (1966) by Barton Maughan; *The Japanese Thrust* (1957) by Lionel Wigmore; *South-West Pacific Area—First Year* (1959) by Dudley McCarthy; *The New Guinea Offensives* (1961) by David Dexter and *The Final Campaigns* (1963) by Gavin Long. Other sources include *The Times History of the Great War*; the British official histories; *Jacka's Mob* (1933) by E. J. Rule; *South Africa—The Transvaal War* by Louis Creswick.

Chapter Twelve first appeared as two articles, by me, in the *Bulletin* in 1958; I have modified them for this book. The short story which makes up the final chapter, also by me, was published in *AM Magazine*, Sydney, in 1953.

Apart from my numerous written sources I have interviewed many soldiers, famous and 'ordinary' throughout my life. I have visited every region in which Diggers have fought and nearly every battlefield. As a member of the 2nd AIF I have, of course, my own recollections and opinions about the Digger.

Once again, I must thank my wife, Hazelle, for her great help in the preparation of this book and for her company in many countries in search of the Digger.

Unsolicited Testimonials

'I cannot surrender ... I am in command of Australians who would cut my throat if I did.'
Colonel Hore, commanding the garrison at Eland's River, Cape Colony in August 1900, when invited by the Boer commander, Delarey, to submit.

'Once again Australian troops showed that the Boers can be confused and even stopped by enterprising men ... The Australians have brought an intellectual appreciation to warfare.'
A Capetown newspaper correspondent writing about the Australian victory at Eland's River.

'Anyhow, I have commanded an Australian division.'
The last words of Major-General Sir William Bridges, mortally wounded at Gallipoli, 1915.

'From the start to finish the Australians distinguished themselves by their endurance and boldness. By their initiative, their fighting spirit, their magnificent ardour, they proved themselves to be shock troops of the first order.'
Marshall Foch, 1918.

'It was a pleasure and an honour to be fighting alongside troops who displayed such magnificent morale.'
Major-General R. L. Mullens, commander 1st British Cavalry Division to General Sir John Monash, Australian commander, 1918.

'The Australian Corps has gained a mastery over the enemy such as has probably not been gained by our troops in any previous period of the war.'
War Diary of British Fourth Army, 1918.

'They are great soldiers, these Australians ... There is a reckless dare-devilry, combined with a spice of cunning, which gives them a place of their own in the Imperial ranks.'
Sir Arthur Conan Doyle, 1918.

'Those damned Australians!'
Field-Marshal Erwin Rommel, 1941.

'The 9th Australian Division struck what history may well proclaim to be the decisive blow at Alamein . . . The magnificent forward drive of the Australians, achieved by bitter, ceaseless fighting, had swung the whole battle in our favour.'
Winston Churchill, The Second World War.

'The 9th Australian Division contributed greatly to the success of Operation Supercharge [the final offensive of Alamein]. In spite of heavy casualties they almost destroyed the 164th German Infantry Division . . .'
Lieutenant-General Sir Brian Horrocks.

'The Australians? They won Alamein for me.'
Field-Marshall Montgomery in conversation with the author, January 1971.

'There is one thought I will cherish above all others—under my command fought the 9th Australian Division.'
Field-Marshal Viscount Alexander, 1943.

'The Aussies are unique soldiers, amazingly casual but willing to tackle anything. They liked to work in small parties and were always successful. If they had any fear they never showed it.'
General Eichelberger, US Army, 1945.

'Man to man and all things equal, the Australians proved themselves worth a score or more of the enemy.'
Reginald Thompson, British war correspondent, referring to the 3rd Battalion, Royal Australian Regiment in Korea, 1950.

'They are the salt of the earth and all those who know them cannot but be inspired by the tremendous job they are doing both for Vietnam and Australia.'
Lieut. General Sir Thomas Daly, referring to the Australian Army Training Team, Vietnam, July 1969.

1

Echoes of a Reputation

Both my parents were Diggers. My father was first a
member of the Medical Corps of the Australian Imperial Force (AIF) and later an officer of the 20th
Infantry Battalion; he served in Egypt, Lemnos (Gallipoli)
and on the Western Front in Belgium and France. My mother
was a sister in the Australian Army Nursing Service and
served in Egypt, Lemnos, on the Western Front and in
England. Much of her service was with the 3rd Australian
General Hospital. They married in England while both were
still serving but under the regulations of those days my
mother had to leave the army on marriage.

With parents such as these my early interest in Australian
military history was almost assured. Born in 1922 and grow-
ing up in the 1920s and 1930s I heard my parents talking
about the war and sat quietly and listened to conversations
when old comrades called at our home. Many of them were
former patients of my mother, for whom they had great
affection. They openly called her 'Little Sister', and she was
indeed a tiny person. Others called her 'Digger Nell'.

Several of these men told me that my mother had 'pulled
them through'. One explained that he still had legs because
my mother had talked a doctor into not amputating them
when he was suffering from trench feet, an appalling affliction caused by standing in icy water and mud for far too
long. She had told the doctor that good nursing could restore
the misshapen, rotting feet. They were clubbed and twisted
in 1930. When they had recovered sufficiently, these particular Diggers used to scrounge for my mother and no doubt
for other sisters. Her medical orderlies—for they too called
at our home—seemed to have made a profession of scrounging; they used to bring her juicy steaks stolen from some

British officers' mess on Lemnos or from one of the ships in harbour. Scrounging was one of the first army words I knew.

My father's old army friends also had a lot to talk about and as he had been in Egypt as well as on the Western Front the yarns were as much about deserts as mud, about 'bloody, thieving Wogs' as well as 'bloody Jerry'. They talked about 'the trenches' in an odd, faraway tone as if recollecting something they couldn't really quite remember. When I was a little older I realised that they remembered very well; their difficulty was in comprehending.

Neither of my parents shielded me from the horrors of their war. They didn't overload me with it or cram it down my throat; but they answered my questions without evasion, even if this meant telling me something which people who had not known the war would have regarded as horrible or horrific. There was another source of information as well—my parents' pictures and postcards of the battlefields and operational areas, unit magazines, newspaper cuttings and their souvenirs. One which intrigued me greatly was my father's webbing belt on to which he had fastened a score of regimental badges swapped or scrounged between 1915 and 1918. In the 1940s I added two badges of my own—a marksman's crossed rifles and a signaller's crossed flags—and the belt is now in my own battlefield museum.

Regularly I went with my father to the Anzac Day march on 25 April; we always arrived in time for the dawn service at the cenotaph in Martin Place, Sydney, where I would try desperately not to cry but invariably did. My mother rarely went to the march but waited at home for her worshippers to visit her. Still wearing their medals, they would troop in during the afternoon to embrace her. One of them, a tall, good-looking man, always bowed gravely and said, 'Salutations, Little Sister.'

She enjoyed these occasions for she understood the men and their language—not that they used *language* in front of her, she checked that. Nor in front of me, my father checked them then.

Army stories fascinated me and Albert Jacka VC, MC was my hero; he was at least on a par with Don Bradman. I might have boyishly regarded him as a superman but I had certainly not then begun to believe that the Australian soldier was anything out of the ordinary, for I was in no position to compare. But as an early and voracious reader, especially about anything to do with the 'Great' War, I found various historians, war correspondents and former soldiers making assessments and comparisons about the Digger.

Sometimes my father took me to a military hospital to visit old friends and one day in 1930 I realised that some of the men he had come to see had been in hospital ever since the war ended—twelve years before. At the age of eight this was a sobering thought and when I mentioned it to my father he said, 'Some of these boys will *never* leave hospital.'

'They don't seem too unhappy about it,' I said.

'Ah well,' my father said, 'that's because they're Diggers.'

One image of fortitude in suffering was particularly vivid— a man who lived in a bath of chemicals because he had lost all his skin during a German gas attack. He lived there until he died and then my father said to my mother, 'I can't remember Roddy ever grumbling.'

'He wouldn't,' she said in a matter-of-fact way.

When my mother wrote her reminiscences as an army nursing sister I typed the manuscript for her—I was fifteen— and learnt even more about the war and Australian soldiers. She wrote dispassionately and well and told of men's pain and suffering; there were men with half a face after a shell-burst, others with wounds so gaping that entrails protruded. What she had witnessed on Lemnos while nursing men brought off the Gallipoli peninsula during 1915 had left a mark on her mind which stayed there until she died at the age of ninety-two. She saw no glory in war—though she recognised heroism; she saw no justification for war—though she understood the motives of the men who wanted to take part in it. More than anything the waste of good men appalled her.

Once, under my cross-examination, she conceded that Australians were good soldiers. 'But what's the point of being a good soldier if you're dead or crippled?' she said. She served soldiers because they were sick or hurt, not because they happened to be good soldiers.

I have said that my parents always answered my questions. Nearly always. Had my father killed a German with a bayonet? Had his friends been killed beside him? These and similar questions neither he nor his friends ever answered. It often seemed to me that when their reminiscing became too serious they deliberately changed it to the cheerfully anecdotal. Then they would talk about a comrade named 'Horrid Harry', who had been told off by a British colonel for impertinence—Harry had addressed the officer as 'mate'—and about the 'hard case' who had scrounged a case of rum. Sometimes, from the conversation of these men, it seemed that the Australian Imperial Force had been a bunch of good-natured larrikins. Much later I understood that the Diggers had become victims of a parody which they themselves had helped to create.

One of the many war histories I read as a boy was *Mr Punch's History of the War*. In a despatch dated 15 December 1915 an English officer was quoted as saying that the Australian soldier was 'the bravest thing God ever made'. This prompted a *Punch* poet to pen these lines:

> *Bravest, where half a world of men*
> *Are brave beyond all earth's rewards,*
> *So stoutly none shall charge again*
> *Till the last breaking of the swords;*
> *Wounded or hale, won home from war,*
> *Or yonder by the Lone Pine laid;*
> *Give him his due for evermore—*
> *'The bravest thing God ever made!'*

Following in my parents' footsteps, as a member of the 2nd AIF, I learnt much about the make-up of the Digger. And I have gone on learning.

Trench maps in hand, I have many times explored the battlefields on which Australians fought in dozens of campaigns and battles. I can locate the exact spot, on the ground, where a certain company of a particular battalion lost many of its men in a grenade fight. I have dug up Diggers' possessions lost in the mud of Flanders. In many towns and villages I have sought recollections of Australian soldiers— in France, Belgium, Egypt, Libya, Lebanon, Israel, Syria, in fact in all the countries where Australians have fought since 1900.

In the 1940s and 1950s, even into the 1960s, first-hand memories of 1914–18 abounded to be replaced, as the older people died, by a folk memory. In a Damascus hotel a Syrian waiter, told that I was Australian, instantly made gestures to indicate a big hat and then performed some bayoneting motions. In the same city a dealer in the bazaar told me how in 1918 Australian horsemen had galloped down The Street Called Straight—that's its name—and laid about them with the flat of bayonets and swords to quell a riot. Both these men were too young to have seen an Australian of that war; they were relating inherited stories and impressions passed on from their fathers. 'The Australians were always kind to children,' a Frenchman in Arras told me in 1970. 'How do you know that?' I asked him. He shrugged and said, 'Everybody knows.'

Sometimes I find a perspective of the Digger in an unexpected place. In my writing travels around the world I am constantly looking for war souvenirs and relics of all kinds— medals, uniforms, weapons, souvenirs made by soldiers in the trenches. This search took me to a quaint old antique shop in the beautiful German city of Heidelberg in the summer of 1957.

Introducing myself to the middle-aged proprietor, I told him my nationality and explained my interest in all things military. 'I would like to talk with you about that,' he said, and excused himself while he attended to some customers. When he returned he asked if I had been in the Australian army during the recent war.

I admitted to this cautiously for sometimes the admission leads to embarrassing reminiscences about Australians being involved in drunken brawls and indiscriminate scrounging.

The antiques dealer, whose name was Karrel, said, 'Then I have something that might interest you; please don't go away.' He hurried into the private section of his shop and returned in a few minutes holding a folder from which he took several letters. 'My son wrote these letters to me during the war,' he said. 'He was one of Rommel's men in the Afrika Korps and fought against the Australians and the British and the South Africans and the Indians and the New Zealanders.'

This seemed like the preamble to an explanation that the Afrika Korps, with the doubtful help of the Italian Army, had little chance of a fair fight against such an array of enemies—although, in fact the Germans and Italians outnumbered the British and Empire forces.

But Mr Karrel had no such thought in mind as he leafed through the letters. They preoccupied him so much that he pushed some browsing American tourists out of the door and locked it. Finally he selected a particular letter and looked at me hopefully. 'You fought at El Alamein?'

I was sorry to have to disappoint him. 'Well, never mind,' he said. 'But my son wrote this letter after Alamein, so I though it would interest you. May I read it to you? But perhaps my English translation is not so good.' So I listened to the letter which the young captain Klaus Karrel had written to his father late in 1942. His father's translation was indeed not so good so we had the letter professionally translated at Heidelberg University next day. The translator, with an idiomatic as well as grammatical knowledge of English, produced a rendering that captures the Australian flavour of Captain Karrel's quotes.

I am in hospital, but do not worry. I have been badly wounded, but I am quite safe, tell mother. I have been in a big battle and the Afrika Korps is smashed, like me. We do not yet know what really happened, but the British bombardment was heavier than I thought shellfire could ever be and the infantry attacks were very fierce and determined. I will tell you more about the battle when I see you, but now I want to recount something very remarkable that happened to me.

I was well forward one night with three men looking to see if the enemy was interfering with the minefields in front of my sector. Then we encountered a small Australian

patrol. You have no idea how the Australians can fight
and they were facing us in this position of the front. I
think there were four Australians. Anyway, they were as
surprised as we were and there was a fight. One of my
men was shot immediately so I grabbed his rifle and tried
to bayonet the man who had shot him. It is very hard to
describe now, for one never knows quite what is happening
at a time like that. But I do know that Australians are
astonishing bayonet fighters and my opponent was too
much for me. I am strong and fit and well trained, but he
overwhelmed me and bayoneted me in the stomach. Perhaps I was lucky, because he was aiming for my throat,
but I deflected his thrust.

Then as suddenly as it had started, it was over. The
Australians had gone and I was lying on the desert, with
my men around me, I crawled to them, but they were all
dead. Only one had been shot, the others bayoneted.

The nights had been fairly quiet, but as usual when
somebody fires a shot it starts everybody firing and bullets
and shells were shrieking over me as I lay there in some
pain and losing blood. I tried to crawl back to my lines,
but movement was too painful. I thought that I would
die, and very badly I wanted a drink of water. I should
have been wearing my water bottle, but I had expected to
be away from my position for only an hour or so. My
party had not been in front of our mines when attacked,
so it was unlikely that men sent out to find us would do
so.

I was frightened, not for the first time, and being helpless adds to one's fear. I lay there for a long time. Then I
heard sounds and soon a man appeared. He was an Australian and I thought as he came close to me that he had
come to kill me. But he said: 'Speak English?'

'Yes,' I said, feeling my revolver. But it had gone.

'How are you feeling?' he said and began to probe my
wound.

I swore at him and all he said was, 'Keep quiet, will
you? If you make a noise I'll knock your bloody head off.'

To my astonishment he began to dress my wound,
cutting away my uniform to do so.

I said: 'What are you doing?'

He said: 'What the hell do you think I'm doing? Do

you want to bleed to death? You have a wound I could drive a car through. For God's sake lie still.'

'Why have you come out here?' I said.

'My mate said he thought you were alive,' he said. 'So I came out to have a look.'

'Your mate?' I said.

He smiled at me. 'The bloke who stuck you.' When he finished dressing the wound he sat by my side. 'I'd give you a cigarette,' he said, 'only someone might see it and use it as a target and I don't want to be hit. Now listen. I'm going back to get a stretcher and then I'll have you carried to one of our dressing stations.'

'No!' I said. 'Don't do that. My men will find me. Thank you for what you have done, but don't take me prisoner.'

'Don't be a silly bastard,' he said. 'If I don't get you to a doctor you'll die before dawn. You could lie out here for days before anyone finds you.'

I told him that I would take a chance on that and begged him again to leave me there. He agreed, saying that it was my funeral. I thanked him again and insisted that he take my watch in payment for his attention. The Australians are said to collect battle souvenirs, but he refused the watch. However, he did search my pockets and took a few things of military value which I foolishly was carrying, and just before he left I made him take my death's-head tiepin.

'You're taking an awful chance, Fritz,' he said. 'I'll be back here with the stretcher-bearers at four o'clock and if your own men haven't found you by then I'm taking you in.'

Just before he left me he did give me a cigarette and some matches. He said if I wanted to risk being fired at I could smoke, but to wait for him to get out of the way. I shielded the cigarette with my hands and smoked it after he had left.

I lay there for a long time. The pain was now severe and I was fainting. Once during the night I opened my eyes to see this Australian on his knees looking down at me. He wet my lips with water, said, 'Hold tight, mate', and then, very suddenly, he made off.

I realised why he did this a few moments later when I

heard sounds from the direction of my lines. I called out and was found by some of our stretcher-bearers. So I was saved, but the doctors told me that if the Australian had not found me and dressed my wound I would have bled to death.

Ever since this happened I have not stopped being amazed. One Australian savagely bayonets me then sends another Australian to save my life. It doesn't make sense. I believe that the Australians when they held Tobruk nearly drove the Field-Marshal [Rommel] mad. Every soldier in the Afrika Korps fears the Australian bayonet and you may be sure that when we know what happened at Alamein it will be found that Australian bayonets defeated us.

The incident has made me very thoughtful and I would like to know more about the Australians. What happened to me is such a contradiction. One man with the hands of a killer, another with hands as gentle as any woman. And yet they are the same man, if you understand me . . .

Klaus Karrel returned from Africa and left the army because of his wound. Later, with Germany under attack from all sides, he and many other men unfit for service were recalled and pushed into combat to check the Russian advance. Karrel died somewhere on the eastern front.

His father asked us many questions about Australia and Australians. His son's experience still intrigued him and he said, 'The Australian soldier seems to be an unusual man. Can you tell me why this is? What sort of man is he, really?'

This book is an exploration of the Digger, an attempt to explain what he is, *really*. Or was. I think he may have disappeared into the jungles and paddy fields of Vietnam, destroyed, perhaps, by the shameful public reaction to Australian sacrifice there. In the new, sophisticated, cosmopolitan Australia, its ethos radically changed by materialism and successive waves of immigrants, I doubt if the Digger can be recreated. But at least he can be remembered.

The Australian 'citizen soldier's' feats at Gallipoli and scores of battlefields after that shaped the nation. Now, I suspect, the nation shapes the soldier. In April 1980 a Melbourne social studies teacher made a survey among his thirty-eight eleventh-year students, aged fifteen and sixteen, to find out what they knew about Anzac. Only nine understood clearly that the word is an acronym for Australia and New Zealand Army Corps. Only six had any idea of the precise location of Gallipoli; many thought it was in Italy. About thirteen had 'no idea' about Anzac. There was a vague feeling that it was somehow connected with 'the war' and one youth dismissed it curtly as 'just another bludge day'.

Anzac and Anzac Day are better understood by the primary school children of the village of Harefield in Middlesex, England. In 1917–18 an Australian army hospital was stationed at Harefield and the staff—which included my mother—and the convalescent soldiers were often seen in the village. The Diggers who died were buried next to the local church.

The villagers adopted the Australians in their midst and have ever since remembered them. On Anzac Day a service is held at the church, which has an Anzac chapel. As a climax to the service the young children, who are taught about Anzac Day at school, reverently place great masses of flowers on the grave of every soldier and army nursing sister and the 'Last Post' is sounded. For Harefield, Anzac Day is the most important day of the year and the unflagging remembrance for seventy years is very moving.

2

In the Beginning

Soldiers of any nation reflect the society and social conditions from which they come. On active service and in battle soldiers project the characteristics and qualities—and the defects—of their home environment. In some armies, coming from countries with a pronounced class and economic structure, active service behaviour can vary greatly from regiment to regiment.

Soldiers themselves, meeting as national allies during some far campaign, are intrigued by the differences they perceive between themselves and troops from other countries. They rarely envy the differences and sometimes they ridicule them.

Increasingly from 1900 war correspondents, historians, other nations' military and political leaders and letter-writers among foreign soldiers came to use a particular set of adjectives and phrases to describe Australian soldiers, as individuals or as an army. Not necessarily in order of frequency of use, they are: independent, casual, ill-disciplined but brave, dare-devil, rough and ready, hard to handle, enterprising, eager to attack, indifferent to rank, unorthodox, chauvinistic, cocky, aggressive, fond of drink and girls, sentimental.

All these terms fit many Diggers at certain times and almost all need qualification. Australian soldiers, unlike those of some other nations, cannot be described in black-and-white terms. From the beginning of Australian military history they were more complex than they seemed to be for they had come from a complex society, historically speaking.

The first Australians were rebels. Whether convicts or free settlers they were opponents of an oppressive social system in Britain. British gaols in 1779 were bulging with 100,000 convicts awaiting transportation to the colonies. That the gaols were overcrowded was hardly surprising. More than

160 offences were punishable by death, ranging from high treason and murder to deer-stealing, cutting down trees, concealing the birth of a bastard child, conspiring to raise wages and 'breaking down the head of a fishpond whereby fish may be lost'. In extenuating circumstances the death penalty was sometimes commuted to transportation with penal servitude, often for life.

Offences not punishable by death made a still longer list. It included grand larceny—defined as stealing goods to the value of more than one shilling. Transportation was the standard penalty. Even soldiers found themselves shipped abroad, perhaps for failing to salute an officer. Three classes of convict were transported to Botany Bay—the 'lifers', the fourteen-year men and the seven-year men. Lifers and 'fourteeners' were destined for hard labour on the roads and public works of the colony while the 'seveners' were assigned as unpaid labourers to private employers. Mary Reibey, aged thirteen, who borrowed a neighbour's pony for a frolic, was one such servant. Benjamin Hall, the son of respectable parents in Bristol, fell madly in love and stole a coloured kerchief, valued at nine pence, to give his girl. His sentence was seven years' transportation—a punishment harsh enough to turn him into one of Australia's first bushrangers.

The free settlers, in a paradoxical sense, had Australian characteristics before they left England. Some were rebels and individuals by choice—protesting against poverty or social injustice or lack of opportunity. Others were rebels because the harsh, oppressive legal system made them so. How could ordinary people feel respect for a country which sentenced a man to hang for stealing a loaf of bread to feed a starving family? These settlers by inner compulsion or legal enforcement were angry, bitter, disillusioned and vengeful.

They took with them from the home country—England, Ireland, Scotland or Wales—the seeds of what was to become the distinctive Australian character as their allies and enemies were to see it on campaign and on active service—enterprise, adaptability and aggressive self-confidence, the demand for a 'fair go' and the attitude that a man was as good as his master. Men and women had to dare in Australia in order to thrive.

They dared the heat, the loneliness of the far country, the poverty and the back-breaking work. As more and more free emigrants arrived there came the pioneering breakout from New South Wales coastal confines. Defying all regulations, the new 'squatters' moved ever outwards, taking what land they could. On the back of their sheep was the future wealth of their country, once a year to roll in bullock wagons to the seaports.

The new colony attracted bankers, burglars and king's officers, scholars from Oxford, Irish parsons and seducing baronets, sprigs of the peerage, masons, iron workers, plasterers and carpenters. Dour Scots reformers, followers of Tom Paine and a small group of aspiring painters, poets and engravers found their way to Australia.

Several hundred convicts sent to Australia were political prisoners. They had been convicted for 'disturbances' during the height of the enclosure movement in England, in which thousands lost their chance of continued employment on the land. At that time, too, new inventions were making the old methods obsolete, forcing men into rebellious redundancy.

Some of the convicts were social outcasts and degenerates but their condition was mainly the result of sudden economic changes which the governments of the time did not understand and handled badly.

Very few of the people who went to Australia had the security of money. The vast majority had to exist by their sweat, their enterprise and ability to improvise, and, often enough, by their wits and cunning. They also found that it was important to have friends, not many of them, but a few on whom they could depend utterly. A man building a homestead or droving sheep, establishing a shop or a smithy, needed—apart from his wife—the mate on whom he could depend.

Many of the Diggers of 1914–18 were removed by only three generations, possibly two, from the hard pioneering days. They were imbued with all the values of their parents and grandparents, but they had a much more tolerant and more mellow attitude towards 'Home', meaning Britain. In the new country the hatred and dislike of the old one weak-

ened and died. By the 1880s there was a great sentiment for
'Home' and when the 'Old Country' appeared to need help
it was readily given.

In a serious way, the Australian military tradition was born
in 1900, became a lusty infant during the South African
(Boer) War and reached adulthood during the Great War.
But active-service Australian soldiers had existed for a long
time before 1900. Among the first were those who enlisted
to fight in the second Maori War, which began in 1860.
Four Australian regiments, a total of 1475 volunteers, were
sent to New Zealand in 1863 and given the title of the
Waikato Militia for no better reason than that most of the
fighting was taking place in the Waikato region. Another
1200 volunteers jointed the Militia in 1864.

On 21 June that year Captain Moor of the 1st Regiment
Waikato Militia led his Australians into battle against the
redoubtable Maoris. It was the first fight for Australian soldiers and the last one of the war. Some Australians stayed
on, under the terms of their engagement, as garrison troops
to protect the settlers from any further Maori uprising.

In January 1885 the New South Wales Government—
there was as yet no federated Australia—offered Britain a
force of 750 infantry and artillery to fight in the Sudan War.
The force was raised, equipped and shipped within two
months, a remarkable feat for a young nation without a
military background.

The men were seen off by an enormous crowd—some
reports say 200,000—who cheered wildly as the marching
column went aboard the troopships *Iberia* and *Australasian*.

The contingent reached Suakin on 29 March and within
twenty-four hours was advancing on Tamii in company with
illustrious regiments—the Grenadier Guards, the Coldstream
Guards, the Scots Guards and Royal Horse Artillery among
others. It was a heady, fiercely hot and gruelling experience
and three men were wounded.

The contingent left Sudan on 13 May and six men died
of tropical illness on the journey home. The party received
a rapturously patriotic welcome and qualified for two campaign medals (not decorations), the Egyptian Medal with bar

Suakin 1885 and the Khedive of Egypt's Bronze Star. This may have been some compensation for the disappointment and frustration which most of the men felt for the lack of real military action.

Still, while the contingent had been given no opportunity to prove itself, it had established a precedent for the colonies to send troops to England's aid in future wars. There was also another important result of the Sudan contingent's service. It brought back to Australia a conviction that the regular British Army was unduly rigid in thought and method, and this conviction helped to give the foundling Australian Army its notable elasticity.

Even so, the Australian attitude towards war was for many years almost entirely British. Australians of the last century *knew* that the British Army was the best in the world and Australians revelled in accounts of English feats of arms. There was a feeling when the Sudan contingent sailed that, after all, these men were really only amateurs and everybody hoped that they would not shame Australia in English eyes. The notion that Australia could not possibly produce first-rate troops persisted for many years.

3

Baptism of Fire

Federation of Australia occurred in the middle of the South African War. All the colonies, and subsequently Australia as a nation, offered contingents. In all, Australia sent 859 officers and 15,064 men to the war—12,000 before Federation and 4000 after. There was a total of fifty-seven contingents made up of fifteen from New South Wales; eight from Victoria; Queensland, nine; South Australia, nine; Western Australia, nine; and Tasmania, seven. The men took with them 16,357 horses and about 220 guns and wagons.

Apart from the notable exception of the NSW Field Ambulance, most of the troops were volunteers of the typical Australian arm, Mounted Infantry, and even a few foot soldiers, sent on the suggestion of the British War Office, were also mounted soon after arriving in South Africa. Campaigning in country very similar to their own, the colonial mounted troops were found more suitable than regular cavalry. This didn't surprise the men, but certainly did surprise the British authorities. The Australian casualties were light, in all 1400 including wounded; 251 were killed in action, 267 died of illness.

The contingents can be divided into three classes: the early ones, paid by local government or by private subscription; the Imperial Bushmen, subsidised by the Imperial War Office; the Commonwealth Contingents, raised after Federation, but also under British pay. Many officers and men of the earlier contingents re-enlisted in later units; many made three trips to South Africa. The important fact is that all were volunteers.

Many writers and observers during the Boer War could see in the Australians (as in the New Zealanders and Canadians) the type of troops England badly needed in campaigns like that of the Transvaal. One of them wrote:

Ex-frontier cavalryman myself, with further experience as
cowboy in both the United States and the north-west
Canada, and also as stockrider in Australia, I have never
for a moment doubted that in the raising of an irregular
Anglo-Boer force lay the solution of England's problem,
'How to successfully cope with the enemy.' Sans standard
of physique, sans much orthodox training, sans everything
but virility, inherent horsemanship, inherent wild-land craft,
mounted on his own pony-bronco of Canada or brumbie
of Australia—the Canadian ranch-hand, the Australian
stockrider, shearer, station rouseabout, or the 'cull' of all
lands anglicised might easily become the quintessence of
a useful and operative force against a semi-guerrilla enemy.
   A pair of cord breeches, a couple of shirts, his big hat,
and a cartridge-filled belt, Winchester carbine, a pony of
the sort that can be run to a white sweat, and staggering,
tremble, and then be kicked out to nuzzle for grass or
die—that's what your man wants. The pants and shirts
will be better than he has worn for years; the gun he has
'shot straight' with, ever since he first handled his 'daddy's'
muzzle-loader; and the 'hoss', why each is of the other,
horse and man; they are apart but a thing complete . . . .
   The theory of weight and height for effective fighting
is exploded. Heart, eye, and seat, and wild-land inherent
tact make up for it. Five-feet-six can ride and shoot and
fight or die as well as six-feet-two. We wild-landers have
proven it over and over again. . . . Your yeomanry won't
do the trick; nor your oat-fed khaki-clad higher Colonials
either. 'Tis your cowboy, stockrider, shearer, rouseabout,
cull, given his way and a cause—yes, he and his scrub-fed
mongrel mount and 'gun'.

This impassioned writer, now unknown, was right, but
British military traditionalism killed many Tommies at
Stormberg on 9 December 1899. An all-English force under
General Gateacre, after a seven-hour trek over rough country,
marched, in fours, right into a Boer ambush. They had no
scouts or skirmishers out and paraded, in mass, into volley
fire from strong enemy positions. Even though the men had

been on continuous duty for about thirty hours, they were not withdrawn into positions which could have been held, but were ordered to attack the strong Boer positions—a frontal attack without cover in broad daylight.

The attack was up hills and finally up a quite unclimbable precipice. From the top of it, under cover, the Boers shot down the English attackers at five metres range.

The English were brave all right, but nobody could accuse them of being intelligent. Why did they act as if they were storming a citadel defended by men with clubs? The whole action was typical of English blunders throughout the Boer campaign, blunders glossed over by such feats as Baden-Powell's magnificent defence of Mafeking and by the frame of mind which holds that there is glory in a suicidal attack.

'Our men,' said Louis Creswick, the war's notable historian, 'many of them having been occupied the whole of the previous day in fatigue work, were numb from exhaustion, dropping here and there, fainting or asleep, in the very face of death.'

Every man of the Royal Irish Rifles and the Northumberland Regiment was killed or wounded at Stormberg; none got back when the order to retreat was given. The retreat was a thirteen-kilometre trek back to a place called Moleno, with the Boers attacking all the time. The troops were kept on the road; apparently nobody thought of scattering them and making them difficult targets. Nobody thought of taking up a defensive position in the rocky country admirably suited to holding out against an attack.

Compare Stormberg with an Australian exploit a few months later.

On 9 February 1900, in Cape Colony, Captain Cameron, commanding the Tasmanian contingent, with Captain Salmon and fifty Australians, started out from Rensburg on a reconnaissance. The enemy was soon encountered and gave the Australians a warm time as they advanced across the plain.

The Australians cleverly took shelter [Creswick wrote] and returned an active fusillade, but the Boers seemed to be everywhere in overwhelming numbers. The Australians with great gallantry took possession of a kopje and held it for a good hour and a half, but the numbers opposing them were too great and when the Boers worked round to the rear and fired on their horses they thought it high time to come down, mount and retire, amid a hurricane of lead. The same action was repeated ... and finally a retirement had to be effected across open plain exposed to fierce volleys from the pursing enemy. Strange to say, very few of the Colonials were wounded, though they held their ground during the day with wondrous pluck and tackled the Boers with dexterity equal to their own.

The only casualty was a horse.

The Australians used ground intelligently, did not make pointless charges, moved back by stages and split into parties which covered one another as they went. It must have seemed strange to Creswick, who was accustomed to English troops being cut to pieces under similar circumstances.

The same day another party of Australians, under a Captain Moor, came across the Boers about to shell a British camp at Slingersfontein. They took up a position on a hill near the Boer guns and the Boers in large numbers attacked the hill, shouting to the Australians to surrender. 'For answer,' Creswick said, 'the Colonials only fixed bayonets and yelled defiance.'

Moor's men held their ground and the Boers could not approach. A sergeant and two men were sent off, through the enemy, to tell the CO of the unit that the party was safe and not to attempt a rescue; that the surrounded troops would get out at night. And this they did. In the whole action they lost one killed—a trooper helping one of three men wounded.

Creswick, an observant writer and a faithful historian, liked the Australians and understood them. He was the first writer to describe them and did so with more perception than the Australian war correspondents present at the Boer War.

It is rumoured [he wrote] that the British officers and those of the irregular troops have not always been in accord. The fact is, that one is a master of discipline and the other a master of independence. The Colonial is accustomed to habits of complete self-reliance; he expects to be treated as an individual and not as a machine. Our military system is a machine-made system and one which, unluckily for us, has been incapable of any of the smart plasticities which warfare with the Boers has demanded. Colonial troops will be led, but won't be driven. They are composed of men of first-rate quality, but not men accustomed to obey orders. The Colonist obeys because of the personal influence of a man or men whom he holds intellectually or morally in esteem, but the word discipline for sheer discipline's sake he is disinclined to understand.

Among the ranks of the Colonials are many men of wealth and influence, men of high character and good education. These could not suddenly be treated in the same way as British regulars, who ... require to be welded together on a system. A tactician once asked the question, What is the difference between an army and a mob? The answer is 'Discipline'. It is discipline that converts a rowdy British youngster into the glorious British Tommy that he is. With the Colonial we already have the trained and independent man, and a system of give and take is the only system that can avert friction between men who, though brothers in blood, have, and always must have, the special idiosyncrasies attendant on their dissimilar forms of life.

After the war Creswick wrote:

The Bushmen were perhaps the most curious and refreshing type of the Imperial Brotherhood. Everyone with an appreciation for the genuine was swift to pronounce them delightful fellows, sound in wind and limb, full of go, spirited and keen for work of any kind that came to hand. In addition to this they were friendly and hospitable, would share their last chunk of bully with anyone suffering from a vacuum and they had the nous to forage for themselves and find their way about in the veldt in a manner that excited as much admiration as surprise. They could ride, too. They sat a buckjumper as a child sits a swing and seemed to be horsemasters as if by instinct.

Australian troops fought in several remarkable actions

in South Africa and none of them won more fame than the Queenslanders.

On 1 January 1900, the Queenslanders took part in an attack on Boer positions at Sunnyside—and astonished British officers and observers. At midday the Queenslanders were steadily approaching the Boer position on three sides. They approached under cover, 'cautious as tigers and nimble as cats', finally firing, and returning the fire but only when they caught glimpses of the enemy. Then they blazed away and continued to approach nearer and nearer till the enemy, in view of the persistent and deadly advance, shrank from his ground and sulkily retired. 'The dexterity of the Queenslanders was remarkable; they stalked the enemy as a sportsman would stalk a deer, criticising their own fire and the fire of the foe with coolness and interest.'

The success of their tactics was complete; the laager was captured and with it forty prisoners. The Queenslanders lost two men, though only one in the advance; the other was shot while rescuing a wounded officer. A third man later died of wounds.

The attack on Sunnyside was not an all-Australian affair. A hundred Canadians, with the New South Wales Ambulance, were involved. The Canadians, on their sector, were as successful as the 200 Queenslanders. This was the first time the Australians and Canadians fought together.

The Sunnyside action should have proved to the British commanders that the Boers could be defeated by intelligent tactics, but the lesson was not learned. British troops, lined up as though on parade, continued to advance in the face of heavy enemy fire. An Australian war correspondent noted that the average English soldier seemed to think there was something shameful in seeking cover and firing from it.

On 22 February 1900, men of the 1st NSW Mounted Rifles were present when British artillery moved up to shell a strong force of Boers. Enemy patrols stopped the guns from advancing, but Lieutenant Dove of the Mounted Rifles, on his own initiative, took out a patrol and by clever use of ground and his men forced back every enemy patrol, allowing the guns to go forward.

Even more important, he saved the lives of many British soldiers, for their officers were on the point of ordering a regiment into yet another suicidal frontal attack in an attempt to clear a path for the guns. The Australians of the Boer War could never see the sense in a charge against enemy machine-guns and rifles when they could take objectives by stealth and cunning.

Late in April, Dove and twenty-five men were sent to find the position of a Boer force and camp and did so without incident. As a result the battle of Houtnek was successfully fought a few days later.

On 28 May 1900 the West Australians were prominent at Klip River, twenty-nine kilometres from Johannesburg. The Boers had barely time to hustle their weapons into the train and steam off as some of the West Australian Mounted Infantry dashed into the station. 'These smart Colonials were very much to the fore all day and showed a vast amount of dash and dexterity.' Major Pilkington and a patrol of thirty were moving in advance of the 11th Division, hoping to find a suitable drift for the passage of troops and guns across the Klip River. The drift was discovered, but also the Boers— a posse of them hovering among the kopjes that flanked the road. The little party spread themselves, rifles in hand, to protect the position they had gained, a position of some importance, since it commanded bridges about two and a half kilometres to east and west of the road. The party divided· into two groups, arranged themselves at each bridge, and endeavoured to make a line—a very thin line—as a uniting link between the groups. Creswick said, '. . . the thirty isolated men deluded the Boers, and caused them to believe that these sturdy defenders of the drifts were supported by a huge force in reserve.'

Early in June Colonel de Lisle's 'sprightly Australians', cutting across country, were chasing Boers and guns almost into Pretoria while the infantry, with sunset, were occupying the coveted positions—were handling the key of Pretoria.

But the Australians, darkness or no darkness, were on the warpath—nothing could stop them. They captured the flying Maxim of the flying Boers, pursued them till they were within rifle fire of the streets—the streets where scurrying and panic-stricken forms were to be seen like ants disturbed running hither and thither. Then Colonel de Lisle, equal to the occasion, profited by the general dismay and the demoralisation to send in an officer under a flag of truce to demand the surrender of the town.

An account of this episode was given by Lieutenant W. W. Russell Watson, a Sydney officer, who was the most prominent actor in the proceedings:

Colonel de Lisle came up, beaming with delight, and said, 'Now lad, you have done so well, are you fit to take the white flag into the city and demand the surrender of the city in the name of Lord Roberts and the British Army?' 'Rather!' said I.

So we tied a handkerchief on to a whip, and after saying good-bye to the others, I started for the capital with the white flag in the air, alone and unarmed.

I had not gone far when I was stopped by an artilleryman, so requested him to take me into town. He did so; but the Landdrost (chief magistrate), the Burgomaster (mayor), and the Commandant-General, were still fighting on in the hills about the city, so the Secretary of State was found, and he conducted me to Commandant-General Botha's private residence. He then telephoned to the Secretary for War, and they then despatched messages to their Generals to come at once to a council of war. First, General Botha himself came; then Generals Meyer and Walthusein and the military governors of the city. By this time I had been there two hours, during which time Mrs Botha kindly gave me coffee and sandwiches, which, as I had not had a square meal for thirty-six hours, were most acceptable.

Now came the discussion of the council. The General asked my mission, and this I told him with as much dignity as I could muster. He looked me up and down, and told me to be seated. They all spoke in Dutch, and some of the Generals were very excited. However, after an hour's chat, they drew up a letter, and Botha informed me that if I would conduct the Governor of the city to Lord Roberts, terms and conditions would be arranged. So they all shook hands with me, and said that I ought to be pleased at meeting their greatest statesmen and Generals. Off I went with the Governor and General Walthusein to Colonel de Lisle, who was waiting on the outskirts of the city for my return. The Colonel then joined us, and away we went to Lord Roberts, who was six miles off; so we did not arrive until 10.45 p.m. He was in bed, so just sat up and said, 'How do you do? If General Botha wishes to discuss with me the unconditional surrender of the town, I will meet him at Colonel de Lisle's camp at 9 a.m. tomorrow. In the meantime, I will not fire a shot. Goodnight!'

So unconditional surrender it was, and that at the cost of little more than seventy killed and wounded.

New South Wales troops were foremost in the Battle of Diamond Hill on 12 June 1900. The NSW Mounted Rifles were ordered to support the 6th Mounted Infantry Battalion and the 1st Western Australian Mounted Infantry was held in reserve. About 2 p.m. the advance began, under cover of pom-pom fire, directed on the Boer guns. As soon as Colonel de Lisle saw that the battalion had gained a footing on the hill he brought the pom-poms under shelter and let go the NSW Mounted Rifles. Leaving their horses in dead ground the Australians attacked—the men being opened out at intervals of about thirty metres. Extended in this way the 350 men of the Corps created the appearance of a much larger force and as they swarmed over the hill with fixed bayonets, the Boers retired to a second position 1100 metres away. 'Darkness was just closing over Diamond Hill when the Boers opened a furious fusillade all down the line, but this was the end, for the British [Australians] had captured the key of the Boer position.' The West Australians, though comparatively few in number, played a major part in the rout which followed Diamond Hill.

Pursuit of the Boers seemed hopeless, but

the Westralians were unappeased. They pushed on to a point whence the Boer army, about 4000 strong, with wagons, guns and cattle, could be seen crossing Bronkher's

Spruit. That place of grievous memories (the English had been trapped and mauled there) awoke its own ghosts, for scarcely had the Boers reached the fatal area than avenging sleet from the magazines of the Westralians brought them to a state of panic. In an instant Boers, wagons and guns were scattering in all directions, while the Colonials, expending 20,000 rounds of ammunition, coolly plied their rifles till the numbers of the enemy were thinned by death, wounds or flight. This was the finishing touch to the battle of Diamond Hill.

Colonel de Lisle of the 2nd Mounted Infantry wrote of his 'high appreciation of the way Captain Antill and the NSW Mounted Rifles advanced to take Diamond Hill yesterday and the gallant way the regiment pushed beyond the crest under murderous fire'.

General Ridley wrote:

My dear Antill, I cannot let you leave the country without telling you how much I have appreciated the services of you and your men. Their gallantry, endurance and cheeriness under great stress have been beyond praise and their skilful handling was admirable.

Brigadier-General E. A. S. Alderson: 'Captain Antill is a dashing and capable leader in action and remarkably cool under fire. I have personally seen him carry out some difficult and dangerous tasks with great success.'

Antill was promoted major before returning to Sydney.

The greatest fight of the Australians in South Africa occurred at Eland's River which, in a much smaller way, was the Gallipoli and the Tobruk of the Boer War.

A small force of Australians, including 'A' Squadron, NSW Citizens Bushmen, and Rhodesians (200 in all), was posted at a crossing on Eland's River to secure the crossing and keep it open.

No sooner had they settled down than they became aware that the Boers were close. Their camp was on a flat plain near a boulder-strewn kopje, enclosed by a girdle of menacing hills which commanded not only them but the nearest point of the river, about eighty metres away. The Australians realised they were in a trap, a trap which must be kept open as long as possible.

With dawn the artillery overture began; 1500 shells were fired.

But these were not fellows to be bombarded with impunity. They examined their resources, looked ruefully at their one gun, a muzzle-loader, which before long jammed. The Boers' fire was too hot and snipers too numerous to make repairs, so they had to wait for night. Then the besieged set to work, with a will, brawny arms and knowing heads to build trenches and shelters, splinter proofs and tunnels which should defy the Boers.

Despite the darkness the Boer artillery continued. The Boers varied the entertainment by firing a pom-pom, a rapid-fire gun that could do a lot of damage. This was too much for the Australians. Lieutenant Annat, a Queenslander, took twenty-five Bushmen into the bush and about half an hour later the pom-pom was silent. Later that night, back in position, Annat was killed by a shell.

The next day, 5 July 1900, the Boer leader, Delarey, sent a messenger to say that two other centres had been taken, that the Boers would shortly be in possession of the entire country and that if the garrison didn't surrender he would blow them off the face of the earth with his 94-pounder.

Colonel Hore, in command of the post, sent back a note saying that even if he personally wanted to surrender, which he did not, he was in command of Colonials who would cut his throat if he did. 'I don't expect that your artillery will change the minds of these men,' he concluded.

The shellfire continued all day. In the evening the Bushmen heard gunfire in the distance and expected relief, but it didn't arrive. A column on the way to their support heard that the post had surrendered and having already given battle to a strong Boer force, withdrew.

The 3000 Boers in the hills confidently expected victory. During daylight they trained their guns on the spot where they knew the garrison must water themselves and their horses at night. After dark they opened fire, but they caught few Australians and Rhodesians, who reached the river by devious routes and watered at almost inaccessible places.

Every night and most days snipers went out into the bush and picked off Boers. Several gun crews were wiped out by night-snipers, who would creep close in the darkness and fire at the gun-flashes. Boers who tried to infiltrate were themselves outplayed at their own game. Forward scouts and listening posts of two or three men killed and routed far stronger Boer patrols. The siege lasted twelve days. Five of the garrison were killed, seven died of wounds, eleven were badly wounded and twenty-seven slightly wounded.

Considering the unremitting attack by six big guns as well as continuous rifle fire these losses were not great. The Bushmen held out through their wits and ingenious construction of trenches, dug deeply and zigzag fashion to minimise the effect of shellbursts.

The hungry, tired garrison was relieved on 16 July by a force under Kitchener, who drove off the Boers. Kitchener is said to have looked over the position and remarked that 'only Colonials could have held out and survived in such impossible circumstances'. A Cape Town newspaper correspondent minced no words about the 'affair at Eland's River'.

Once again Australian troops, this time supported by Rhodesians, showed that the Boer can be confused and even stopped by enterprising men. Outnumbered by more than ten to one and completely outgunned, the Colonials must have enraged the Boers who know from experience that it is easy to overwhelm small British garrisons. The Australians have brought an intellectual appreciation to warfare.

High praise came from a Boer historian in 1905 when he made this comment about the 'Eland's River affair':

For the first time in the war we were fighting men who used our own tactics against us. They were Australian volunteers and though small in number we could not take their positions. They were the only troops who could scout into our lines at night and kill our sentries, while killing or capturing our scouts. Our men admitted that the Australians were formidable opponents and far more dangerous than any British troops.

Until recent times the stand at Eland's River was practically unknown in Australia. Even C. E. W. Bean, that most able and thorough historian, while briefly mentioning Eland's River in one of his books, implies that it was not a major action. However, it was a major action in Australian military history, since it established not only the Australian's ability to stick it out under great difficulties, but also his resilience and will to fight back.

Bean leaves one with the impression that Australians saw little action in South Africa but in fact Australians saw much action in the Boer War. The history of the 1st Australian Horse (not to be confused with the 1st Commonwealth Horse) was a typical unit.

The 1st Australian Horse was a volunteer bush cavalry raised in 1895 for service in the remote districts of New South Wales. It comprised several squadrons and its members were all men of high physical and mental standard. They wore a distinctive uniform of dark green with black embroidery, in the hussar pattern, with attractive belts and accoutrements, including sabretaches. Between 13 January and 5 September 1900, the unit took part in thirty-eight actions. The unit's sergeant-major, Griffin, was the first Australian to be killed in the war.

The unit's terse official report includes, briefly, the story of Trooper Palmer who was shot in the head during a fight at Poplar Grove. The trooper bandaged his own head and rejoined his troop, until loss of blood forced him out of action.

The early Diggers (not then called Diggers) were just as tough as their successors and their Boer War records bristle with arduous marches and actions. Lieutenant Sweetland, commanding a section of Royal Australian Artillery, marched 210 kilometres in six days, over very bad roads, bringing all his horses and men in fit and well. The 2nd NSW Mounted Rifles, in ten months, travelled about 6500 kilometres in every part of the Transvaal, fighting and working hard almost every day. In one night march the unit covered seventy-three kilometres, including a difficult river crossing.

In February 1900 Australians took part in the relief of Kimberley. Creswick commented that the relievers of Kimberley looked sorrier specimens of humanity than the relieved, except for the Queensland and New Zealand contingents and the NSW Lancers 'who, considering all things, were wonderfully fit after playing a conspicuous part in the proceedings'. The Lancers rode on the extreme right of the first brigade.

The Lancers played a big part throughout the South African campaigns and were, in fact, the first unit to reach South Africa, although Queensland was the first state to send volunteers. Under Captain C. Cox, a Lancers contingent was in London for a military tournament and training when war broke out. They volunteered to fight and seventy-one of them reached Cape Town on 2 November 1899.

The Lancers were the senior cavalry regiment of New South Wales, having been raised in 1883 as Light Horse to fight in the Sudan. When the men returned to Australia in 1885 it was converted into the Lancers, as a compliment to the 5th Royal Irish Lancers, who camped with the NSW Artillery at Handoub.

The Lancers were probably the first to establish the tradition of Australians always being the spearhead in an advance or attack. They took part in heavy fighting at Klip River on 24 May 1900, and their commander at the time, Major Lee, was complimented by General French on the squadron's work as advance guard.

On 6 June 1901, about 100 South Australians of the Fifth Imperial contingent made an early morning march and captured the whole of De Wet's convoy carrying six months' supplies. For four hours the South Australians, mostly from the Mount Gambier and Naracoorte districts, with another 100 mounted infantry, held out against the famous Boer leader and 400 Boers who tried to recapture the convoy of wagons. This was a great blow to the Boers and a spectacular triumph for the Australians, who made the foray on the orders of their own commander, Captain J. R. B. O'Sullivan. In a year's fighting the men did not spend three consecutive nights in any one place. The 5th and 6th South Australian contingents trekked 6156 kilometres, but still found time to catch and break in 867 veldt ponies as remounts.

Another notable operation was at Valkheuvel Poort on 3 June 1900, when General French again personally thanked the men for their gallant conduct. In all, the Lancers were in action at no fewer than thirty-three places.

Many individual Australians won glory in the field and five returned to Australia with a Victoria Cross. One of the most successful Australians, however, won no decoration at all. His name is now unknown in Australia but it deserves to be enshrined on the national consciousness. For this Australian, Lieutenant Grieve, a Special Service officer, established for all time the character of the Digger officer.

Addressing the Australian nation, Major-General H. A. Macdonald, commanding the Highland Brigade to which Grieve was attached, wrote on 19 October 1900:

> You should all be proud of Lieutenant Grieve. He was ever foremost in his desire to show that those from Australia would not be behind in the attack; and thus he was always ahead, gallantly leading his men to victory, men who deplore his loss and knew his skill as a leader. The officers and men of the famous battalion to which he was attached [2nd Black Watch] mourn for him as for one of themselves, and in the history of the Royal Highlanders, the name of Lieutenant Grieve will find honourable mention and abiding record, an honour to his people and to his country.

Grieve's CO noted that the lieutenant had 'much influence with the men, in whose welfare he took great interest'. This was the secret of Lieutenant Grieve's success and remains the reason why Australian officers ever since have had such remarkable influence over their men. Grieve was killed in action at Paardeberg Drift on 18 February 1900.

On 29 May 1900, Lieutenant Kirby with a company of the 2nd Victorian Mounted Rifles 'bought into' a fight at a railway station, captured several engines, a great deal of rolling-stock and an ambulance train. He was awarded the DSO.

Another DSO winner was Captain Tivey, who on 11 February with forty men of C Squadron, 4th Victorian Regiment, made a forced march of about sixty-five kilometres to Philipstown and surprised more than 300 Boers, drove them back and kept them back until reinforced. The action prevented the Boers from occupying a key position which would have dominated Philipstown.

All five VCs won by Australians during the Boer War were awarded for feats of bravery involving rescue of wounded comrades under fire.

Lieutenant F. W. Bell of the 5th West Australian contingent (Mounted Infantry) won his VC at Brakpan on 16 May 1901. When retiring through heavy fire after holding the right flank, Bell saw a man dismounted and returned and took him up behind him. The horse could not take the weight, so Bell remained behind and covered the trooper's retreat until he was out of danger; then he calmly extricated himself.

On 1 September 1900, two Tasmanians won the VC. A party of twenty men of the First Tasmanian Imperial contingent under Lieutenant G. G. E. Wylly went out after cattle, but were surrounded by Boers. Although wounded, Lieutenant Wylly, seeing that one of his men was badly wounded in the leg and that his horse had been shot, went back to the man's assistance, ordered the man to take his own horse and opened fire from behind a rock to cover the retreat of the others, at the imminent risk of being cut off himself. Colonel T. E. Hickman, DSO, considered that Wylly's conduct saved Corporal Brown from being killed or captured and that his firing to cover the retreat saved other men from death or capture.

The second VC of that day was won by Private Bisdee, one of an advance scouting party. Passing through a rocky defile in the Transvaal the party was ambushed and six of the eight men were hit, including the two officers. The horse of one of the wounded officers broke away and bolted. Private Bisdee gave the officer his stirrup leather to help him out of the action, but finding that the officer was too badly wounded he placed him on the horse, mounted behind him and took the officer to safety. Bisdee was under heavy fire all the time.

Lieutenant Maygar of the 5th Victorian Mounted Rifles won the VC at Geelhoutboom on 23 November 1901. He galloped out and ordered the men of a detached post, which was being outflanked, to retire. One of the men had his horse shot from under him when the Boers were almost within 200 metres. Maygar dismounted and lifted the man on to his own horse, which bolted into a bog. Maygar extricated the horse, told the trooper to gallop for cover, while he himself returned on foot. The whole incident took place under heavy fire.

At Vredesort in July 1900 Captain Neville Howse of the NSW Medical Staff Corps picked up a wounded man under very heavy crossfire and carried him to safety. For this he was awarded a VC.

One of the most outstanding acts of gallantry in the Boer war was performed by a sixteen-year-old bugler named Forbes of the 3rd Queensland Mounted Infantry. Bugler Forbes took his horse and that of his OC, Captain Echlin, behind a deserted farm-house. Both horses were shot and a bullet hit the bugler's haversack. He took shelter with other horse-holders in the farm-house. Ammunition was running desperately low, so Forbes went out under heavy fire and ransacked the saddle-bags on the dead horses for more ammunition. He won the DCM.

Considering the number of their engagements—they ran into many hundreds—the Australian casualties were remarkably light. The greatest number—ninety—occurred in a single action—the defence of the guns at Brakenlaagte on 30 October 1901. Here the Australians were under direct English command.

The Australians in the Brakenlaagte action were Victorians, specially recruited in Australia for the Marquis of Tullibardine's Scottish Horse. Ninety-six of them took part in the defence of the guns and held out until only six were left unwounded. Five officers and twenty-eight men were killed. The Marquis said later that he had never heard of better or more determined fighting. Many more of the Victorians would have lived had they been able to fight the action their own way.

Not only the fighting Australians won acclaim in South Africa. One of the most talked-about Australian units was the NSW Army Medical Corps whose army hospitals and field ambulances were active throughout the campaign. The Field Ambulance quickly won the reputation of being first on the scene at any battle and its men pressed forward under fire to bring back wounded men.

Historian Creswick noted that the NSW Ambulance Corps kept up with the column which relieved Kimberley and was complimented on being the first ambulance to cross the Modder River. The day the corps pitched camp in South Africa it was sent to Orange River on 7 January 1900, and attached to the field army operating under Sir Redvers Buller. The principal medical officer, Colonel W. D. C. Williams, noted that 'this was the highest possible post of honour'.

Within two weeks the corps was handling double the number of cases planned for it. Small sub-units were going off daily to join British columns on the march and in action. At one time Colonel Williams and Major Fiaschi were running the hospital by themselves—with twenty cases of dysentery, twelve of enteric fever and a large number of other cases.

The Commander-in-Chief in South Africa, Lord Roberts, recommended the NSW Field Ambulance as a pattern for the British Army.

Australian nursing sisters accompanied the troops to the Boer War, just as they have been with them in nearly every campaign in the years since. The girls of the Medical Corps of 1900 were capable and uncomplaining, gentle and compassionate.

A forgotten soldier–diarist of the war, in a letter home, said:

From now and for ever I am in love with all Army nurses. I was brought in yesterday, wounded and feeling frightened, and the first person I saw was a Victorian nurse. She smiled at me and said: 'Well soldier, I'll do what I can to help you but you'll have to look a bit more cheerful.'

I was cheerful from that moment. Her name is Sister Rawson and she has promised to see me when we are both back home.

Sister Rawson, one of the ten Victorian nurses, was awarded the Royal Red Cross. Sister E. Binsmead of South Australia and Sister E. Nixon of New South Wales also won the award. New South Wales sent fourteen nurses and South Australia at least three. Others probably served but records were either not kept or were lost.

The Boer War also established the Australian Army padre. Chaplains accompanied most of the Australian contingents, at a time when comparatively few British units had them. His experience of Australian padres prompted Lord Roberts, after the war, to suggest that a chaplain be attached to every British unit. 'I noticed,' he said, 'that the chaplains attached to the various Australian forces were much respected and had great influence over the men. They helped materially in keeping Australian morale as high as it was.'

One of the earliest padres to make his name was the Rev. E. C. Beck, chaplain with A Battery, Royal Australian Artillery. The artillery commander, Colonel S. C. Smith wrote:

Today [21 November 1900] the Rev. Beck left to return to Sydney. His loss is much felt by the battery. By his kindness and forethought, by his interest in and assiduous attendance on the sick in the hospitals he has endeared himself to all ranks, both military and civil. His services were invaluable, particularly to the wounded after the battle of Keis and also to the sick and dying at Draghoender. He has carried out his duties quietly, conscientiously and unostentatiously.

When the Australians finally returned to Australia in 1902 they had won a remarkable reputation for dash and skill in action—and even then they also had a reputation for some wildness and irresponsibility when not in the line.

But they found a nation not particularly appreciative of what they had done, not fully conscious of the unusual and unorthodox character of this new being, the Australian soldier. Their comparatively few casualties were actually held against them. The general attitude was: 'How could the Australian units have seen much action when so few were killed?' This attitude was natural enough at the time, for Australians were accustomed to heavy casualties whenever the British Army went into action. They did not realise that the Australians, unit for unit, accomplished more than any British regiment but at far less cost in killed and wounded.

British generals were honest and generous in their praise and many of them wrote to Australian newspapers in an effort to make the deeds of the Australian troops more widely known, but there is little evidence to show that the public was impressed. Australian war correspondents, who throughout the war were always close to the front-line troops, were equally unsuccessful in making the public realise what the soldiers had accomplished.

Nevertheless, the Australians were by now highly regarded as fighters by senior British officers and by the British rank and file, who were astonished at what they considered the informal unmilitary relationship between Australian officers and men. They were appalled to find that, in action, private soldiers often called officers by their Christian names.

A Queensland captain was sharply rebuked by an English colonel for telling his men details of a projected attack. The colonel said testily that it was not done for officers to discuss battle details with private soldiers. The captain, whose name is unknown, said: 'I don't regard them as private soldiers, sir, they are my mates. Naturally, I want them to know why I'm asking them to risk their lives.'

For their part, the Australians were astounded to see British battalions mass for an attack, instead of dispersing. They saw one unit march into the mouths of spitting Maxims while the regimental drummer beat step. They saw heroic officers armed only with swords lead impossible frontal attacks when a little more appreciation of the region would have shown them a much better way of succeeding. They came back from the war with immense respect for the bravery of the British Army, but none whatever for its intelligence.

And, most important of all, they returned with a sublime confidence in themselves and their officers. They had been tested and in no way had been found wanting. They had planted the seeds of a legend.

One event in South Africa soured the success of the Australians. This was the fate of Lieut. Harry Morant, 'The Breaker' —he was a horse-breaker—who went to South Africa with the 2nd Contingent of the South Australian Mounted Rifles. After this tour of duty Morant went to England but returned to the Cape to join the Bush Veldt Carbineers. While serving with this unit he is alleged to have murdered Boer prisoners. A blatantly unsound courtmartial condemned him to death for this 'crime' and he was executed by firing squad on 27 February 1902. The Australian Government thought that General Kitchener's conduct in forcing through this courtmartial and sentence without reference to Australia was high-handed and arbitrary. As a direct result future governments made it clear that no Australian serviceman could be sentenced to the death penalty by a British courtmartial.

4

Prelude to Battle

When World War I broke out on 4 August 1914, Australia was in the middle of a domestic crisis, for on 30 July the Governor-General had dissolved both Houses of Parliament, following rejection of two bills which the Cook Government was trying to force through.

Despite their political differences the Prime Minister, Joseph Cook, and the leader of the Opposition, Andrew Fisher, agreed on Australia's duty. On 31 July Cook had said: 'Whatever happens, Australia is a part of the Empire and is in the Empire to the full; when the Empire is at war, Australia is at war.'

Fisher, equally emphatic, said: 'Should the worst happen after everything has been done that honour will permit, we Australians will help and defend the mother country to our last man and our last shilling.' Before the war was over Australia had nearly spent its last man and shilling.

When the electors went to the polls on 5 September they knew where both parties stood. Andrew Fisher became Prime Minister, for the third time, on 14 September and, as somebody said, 'got busy with the war'.

Meanwhile the Inspector-General of the Australian Military Forces, Sir William Bridges, had, with his staff, worked out in the first four days of the war details for the organisation of the Australian Imperial Force, so named by Bridges himself.

The call for volunteers evoked an incredible response. Men came from all parts of the commonwealth, this being the first time in history that the states had acted in concert.

There was much national emotion, expressed in the press and at rallies. Wherever Australians gathered they sang the popular song of the day, written by W. W. Francis:

On land or sea, wherever you be,
Keep your eye on Germany!
For England home and beauty
Have no cause to fear!
Should auld acquaintance be forgot?
No! No! No! No! Australia will be there!
Australia will be there!

When enlistments started on 10 August 1914, most of the men who flocked to the barrack squares of the main Australian cities wanted to enlist in the mounted arms, in the Light Horse, Horse Transport and Artillery. Australians at that time had no fondness for walking. Even so, many fine horsemen joined the infantry rather then be left out altogether. So keen were some men to enlist that they broke down and wept when rejected. Nominally, the army required that men be 5 feet 6 inches (168 cm) or over, have a 34-inch (86 cm) chest, be aged between nineteen and thirty-eight and have most of their own teeth. There is the case of one man from western New South Wales who tried to enlist on eleven separate occasions. On the twelfth he finally made it—but only because the examining doctor was his brother. Early in the war the medical and dental standards were extremely vigorous. After a few months men who had been rejected for defects of teeth, eyesight, feet and other things were readily accepted.

The great fear of being rejected as medically unfit was equalled by another anxiety—that the war would be over before the Australians could get into it.

The government's resources were strained to the utmost to produce equipment, uniforms and all the great variety and quantity of items needed for a modern army. But in little more than a month, with a rapidity never equalled during World War II, an army nearly as large as that commanded by Wellington at Waterloo was completely fitted out for service, not only with the personal accoutrements and uniform of the troops but with wagons and harness, medical equipment and all the hundreds of things that go to make up a self-contained force. Liners and freighters were refitted to carry troops and horses; the ships were provisioned and ready to sail. This was an extraordinary achievement for a nation whose only military background virtually had been the Boer War. By mid-September the newly formed brigades and battalions were sufficiently trained to march through the streets, but not all citizens were highly impressed. The loose, peasoup-coloured, dull-buttoned khaki uniforms of the AIF, though more workman-like than those of any other army within the AIF's subsequent experience, appeared slovenly to critics, who compared them with the traditional tight-fitting goldbraided brass-buttoned uniforms of peace-time.

Unit colour-patches were authorised, two-coloured patches of cloth stitched below the shoulder line of a tunic. The shape indicated the division, the upper colour the battalion, and the lower colour the brigade; thus the rectangular black over green was the colour patch of the first battalion of the first brigade of the first division. Men quickly became proud of their colour patches, which had a binding quality in a battalion. The only object of which the Australian was more proud was his 'rising sun' hat badge and, later the AUSTRALIA metal badge which soldiers wore at the shoulder end of their epaulettes.

There was an unevenness about the troops that had people saying: 'They'll never make soldiers out of this lot. The light horse may be all right, but they have the rag, tag and bobtail of Australia in the infantry.' C. E. W. Bean quotes the manager of a big Sydney newspaper who said: 'Going to the front! They'll keep the trained British regular army for the front. The nearest these will get to it will be the line of communications.'

As it happened, during World War I almost every Australian soldier was a front-line soldier. All supply and back-area duties were carried out by British units. The pay of 6s a day for a private, of which 1s a day was deferred to be paid on discharge, was the highest of any army in World War I, and was a fair wage, since it was as high as the wage of the average Australian worker. It was at this time the troops were called 'Six-bob-a-day tourists'. Many people supposed that the main motive of the men enlisting in the AIF was to see the world and not to fight.

Boer War contingents had been better paid. Buglers and privates received 2s 3d a day with another 2s 3d deferred pay; corporals 4s 9d and 2s 3d; sergeants 5s 9d and 2s 3d; lieutenants 16s and 3s; captains £1 and 3s 6d. Officers and men of the Permanent Forces received, after landing, Imperial rates of pay in addition to these rates, so that while serving abroad a private earned 7s 3d a day.

Meanwhile, Australians had already been in action. Within about a month after the declaration of war Australian and New Zealand ships had lowered the German flag in every one of the German possessions in the Pacific. This was a surprise to the Germans, for its military and political writers had said many times that if a great war occurred Australia would declare its independence and set up a republic. On 31 August, Samoa was surrendered to HMAS *Australia*. Early in September the Union Jack was hoisted at Rabaul, the capital of German New Guinea (Kaiser Wilhelm's Land) and at Herbertshohe, the administrative centre of the Bismarck Archipelago.

It is wrongly supposed that the first casualties of the AIF occurred during the brief New Guinea–New Britain campaigns. But in fact these men were not part of the AIF but the ANMEF. Nevertheless it was here that the Germans first realised that Australians were men to be reckoned with. Among their other exploits was the destruction of the great German wireless mast at Bitapaka. The German commander of the post later reported to his governor that the defenders had underestimated the skill of the Australians in jungle fighting; it had been expected that they would be confined to the roads. This was not the first time the enemy found that Australians were never confined to the beaten track, nor the first time they were underestimated.

Although the AIF was ready to go to war in the last week of September it was not until 24 October that the ships began to assemble in King George's Sound, the great harbour of Albany. On 1 November the *Orvieto*, carrying General Bridges, his staff and 1000 Victorians, led the convoy out of the sound.

General Bridges was one of the first really great Australian

New South Wales Lancers on parade in the 1890s.

Harry 'the Breaker' Morant, court martialled by the British for allegedly killing Boer prisoners of war and executed by firing squad in Pretoria on 27 February 1902.

Lieut.-Colonel James Burns of the NSW Lancers, one of the best known soldiers of his day, in 1899.

This photograph is inscribed on the back 'Six mates in Egypt in 1916'. To be photographed in front of the Sphinx was almost a ritual for AIF men.

A Digger gives a drink to a wounded Turk at Gallipoli. (*Australian War Memorial*)

Mates. This group of Army Medical Corps men had themselves photographed at a studio in Cairo. The author's father is on the right.

Because of Simpson's missions of mercy in bringing wounded from the firing line he appears on the Anzac Medallion, struck in 1967. (*Australian War Memorial*)

soldiers. As C. E. W. Bean said, he was a deep student of his profession, a man of first-rate brain and wide learning, with great physical and moral courage and with grim determination.

Bridges, with the help of Major (later General) Brudenell White, was responsible for the formation of the first AIF. It was established very early that the Australian force should be a compact self-contained one, not be split up among other units of the British Army. Australia has reason to be glad of this decision.

The first convoy comprised twenty-six Australian transports, ten New Zealand transports and four warships, the British cruiser *Minotaur*, the Japanese cruiser *Ibuki*, and the Australian cruisers *Melbourne* and *Sydney*. Two days later two West Australian transports met the fleet at sea, making in all thirty-eight transports and four warships, one of the largest convoys ever to leave Australia in any war.

Despite his ruthlessness—and the Australian soldier *is* ruthless when in action—he is still a humanitarian. He has proved this many times. The first occasion during the Great War was on 15 November 1914, when the victorious *Sydney* rejoined the convoy at Colombo after disposing of the German raider *Emden*. The commander, Captain Glossop, had telegraphed ahead a special request that out of consideration for the sixty-five maimed and dying Germans he had aboard the *Sydney* there should be no cheering when his ship passed through the convoy. There was none. Even so, a verse was added to 'Australia Will Be There':

> You've heard about the Emden *that was*
> *cruising all around*
> *It was sinking British shipping where e'er*
> *it could be found,*
> *Till one fine summer morning Australia's*
> *answer came,*
> *The good ship* Sydney *hove in sight and*
> *put the foe to shame.*

The *Emden* could have caused tremendous damage had it

got loose in the convoy. For two months before the convoy left, the *Emden* had been on the rampage—sinking or capturing seventeen ships. A few days before the AIF left King George's Sound, the *Emden*, disguised with a dummy funnel and a neutral flag, had raced into Penang Roads, torpedoed a French destroyer and a Russian cruiser and disappeared.

Originally the Australians were supposed to go to England but great difficulty had been experienced in trying to build winter camps even for British troops and the Canadian contingent already there. The campsites were deep in mud, and training was hampered by these unexpected miseries and by the discontent that resulted.

The Australian High Commissioner in London, Sir George Reid, fearing the effect of this on the Australians—unused to such conditions—asked Lord Kitchener, then the British Secretary of State for War, to keep the Australians in Egypt. The decision reached the convoy only just in time.

By mid-December the 1st Division's training was in full swing in the desert. There was then no danger of air attack and camps were set up neatly and symmetrically. This was the first time many of the Australians had seen a division in camp with the tents of the Division's headquarters, three infantry brigades, artillery, transport and ambulances. It was here that the Australians first made the acquaintance of the Arabs, not only in Cairo and around the pyramids but on spaces near the tent lines where many Arab and other tents were put up, with such labels as 'Australian-style Afternoon Tea'. Other tents were tenanted by tobacconists, tailors, photographers, souvenir-sellers, hairdressers and newspaper-sellers.

Although many of the officers were themselves inexperienced, the training of the troops was efficient and vigorous, carried out as it was under the orders and guidance of General Bridges and Colonel Brudenell White. It was during this early stay in Egypt that absence of leave became common. But most of the offenders were old soldiers who had served in the South African War and were still unsettled in habits. The New Zealand force found a remedy by returning the more troublesome men to New Zealand and soon after this General Bridges inflicted the same punishment on miscreant

Australians. For nearly two years this remained the supreme punishment in the AIF.

In December 1914 Lord Kitchener decided, with the agreement of the Australian and New Zealand Governments, to form Australian and New Zealand contingents into an improvised army corps. To command this corps he chose William Riddell Birdwood, a British cavalry officer, short and dapper in figure, vigorous, brave and an outstandingly good fighting man. He arrived in Egypt from India with his staff on 21 December and organised the corps as follows:

Corps headquarters: A British unit, the staff officers selected by Birdwood in India

First Australian Division

New Zealand and Australian Division (this second division, usually known as the NZ and A Division, would be under the command of General Godley and would comprise a New Zealand infantry brigade, 4th Australian Infantry Brigade under Colonel Monash which was then just leaving Australia)

First Australian Light Horse Brigade

NZ Mounted Rifles Brigade (the only artillery of this division was a single New Zealand field artillery brigade).

The Australian Government should have insisted that at least a proportion of the staff officers were Australians. It was bad enough to import a British general to command the corps. Birdwood was popular enough, but many Diggers objected to being 'shoved around' by any English officer. A quota of Australian staff officers would have brought about much closer feeling between Corps HQ and the rank and file. English officers were rarely able to 'handle' Australian troops.

The men of the AIF left Australia as heroes but saints they were not. They were normal men with a healthy interest in women and all too often the women were unhealthy. Some men were taken off the first convoy because they were suffering from venereal disease. It became the greatest problem for the medical services, which in Egypt established a quarantine camp at Mena, near Cairo. Here VD sufferers were treated in forced isolation and were required to wear a white band on the right arm as an indication of their affliction.

The vast majority of officers and men knew nothing about sex hygiene and the effects of venereal infection. The sufferers were young and they cheerfully believed that VD was easily cured. They commonly regarded continence not only as harmful to the health but unmanly. More than most soldiers, the Australians were conscious of their virility and the need to prove it by frequent intercourse. The man known to be a virgin was sometimes laughed at in a mock-pathetic sympathy, more frequently he was derided. 'What's the matter, mate, can't you get it up?' his comrades would say.

Officially VD was a 'wilfully contracted disease', really a self-inflicted wound, and therefore punishable. A man could lose his pay and his paybook was stamped with his shame— 'VD'. As worried officers pointed out to the men, they could lose the respect of their families.

It was lust which led to probably the most famous or infamous brothel battle in history. Cairo's red light area was the Haret el Wasser near Shepheard's Hotel—a street of brothels known to the troops as 'the Wazzir'. Many of them had grievances about their treatment in the Wazzir. For one thing, the price had gone up, a natural result of supply and demand as more and more troops reached Egypt. Some men said they had been robbed 'while their backs were turned', or that they had been served cheap drink when they had paid for good wine. In addition, they felt aggrieved by the large number of their mates in the VD quarantine camp.

On 2 April 1915, knowing that next day they were to leave Egypt for Gallipoli, men of the 1st Division raided the Wazzir. They destructively sacked the eight-storey houses, threw the bedding and girls' dresses down to the street and set fire to everything. The arrival of the British military police led to a bloody 'stoush' and when the Cairo fire brigade tried to put out fires, the troops cut their hoses to pieces.

In June the troops of the 2nd Australian Division fought the Second Battle of the Wazzir on the same lines as the first. Having enjoyed the pleasures of the Wazzir the men plundered it. It was perhaps a way of expiating a guilt complex but the Diggers would not have seen it that way. They needed sex as a simple but compelling physical necessity and no degree of exhortation, advice or punishment kept them from it. Only battle.

5

Bullet, Bayonet and Bravery

The New Zealanders were in action earlier than the Australians, for at dawn on 3 February 1915, the Indian sentries on the Suez Canal saw several boats being launched by Turkish troops on the opposite bank. Only one reached the western bank and that like the others was riddled with fire from the Indians and a detachment of New Zealanders. No Australian infantry was engaged in this fight, but immediately after it two battalions, the 7th and the 8th, were moved from Mena to the canal and temporarily put into trenches there. By this time the Australians were becoming eager to get into battle.

On 28 February on sudden orders, the 3rd Australian Infantry Brigade with some engineers, ambulance men and some other troops, disappeared from Mena camp. The destination was the island of Lemnos and the move was the prelude to Gallipoli.

On 1 April all leave was stopped. On 3 April camp was struck. The infantry, artillery, engineers, ambulances—all went off to Alexandria to join ship, leaving only the camps of the Light Horse and Mounted Rifles watching forlornly the departure of their comrades (though the Light Horsemen themselves were to make history of their own very soon).

The Gallipoli Peninsula guarded the way through the Dardanelles. The Allied High Command wanted to force the Dardanelles, take Constantinople from the Turks, cut off Turkey from the Germanic powers and, by opening the Black Sea, enable easy communication between Russia and the West. This was the reason for the Gallipoli campaign, the reason the Anzacs left Lemnos for the landing on the beach near Gaba Tepe.

The word Anzac was coined some weeks before the troops left Egypt. C. E. W. Bean suggests that the originator of the word was Lieutenant A. T. White, but claims have been made on behalf of Sir Ian Hamilton, General Birdwood and Major C. M. Wagstaff, among others. The most likely originator seems to have been a now unknown clerk who saw the letters 'Australian and New Zealand Army Corps' stencilled on cases of supplies and realised that Anzac would be a useful word for telegraphic purposes. With Birdwood's approval it was adopted, but as the clerk was probably a private his suggestion must necessarily have gone through officers to reach Birdwood.

The word had a pleasant ring and it came easily to the tongue, but nobody had any intention of perpetuating it. It came into army orders and writings only slowly. But it became as much a part of the Australian tongue as *Digger*.

According to Bean, Digger became common among Australian and New Zealand soldiers in 1917, in France. He says the word was evolved from the gum-diggers of New Zealand. My mother said that she heard soldiers using the word as a form of address in 1916. Most old soldiers believe that Digger came about as a natural result of their trench-digging activities in France, while others have told me that some West Australian soldiers, gold-miners in civilian life, started the word on its way. At one time Digger was a slang expression for a plodder, an apt term for an infantryman. No one explanation is commonly accepted, but many Great War Diggers have contested Bean's explanation.

Tremendous national pride was fanned into flame among the hills and gullies of Gallipoli. The Australians, born exhibitionists, wanted to prove to themselves and to everybody else that they could do it. This was the spirit that bred love of bayonet fighting, disregard of danger and the desire to attack. No individual soldier, from private to field-marshal, can be a good soldier unless he has the desire to attack and no army can be great unless its troops are of this type.

This is the great difference between Australian troops and those of some other countries. The Australian *wants* to go in and glories in close-quarter fighting. Others go in because they are told to do so and have no feeling of glory in their task. This may not make them any the less brave, but it does mean they do not have the supreme will. They fight, but there is an inner, subconscious hanging-back. They don't enjoy fighting and as for close-quarter fighting . . . why go out of your way to find *that*?

The Australian has no inner rejection of a coming combat. He often wants to go in so eagerly that, as many junior commanders know, he must often be restrained from making an enthusiastic but ill-advised attack. The Australian temperament excelled at Gallipoli, for here the conditions as well as the enemy were a constant challenge, giving the Digger plenty of scope for improvisation, adaptation and initiative.

The AIF came ashore on a bullet-swept beach and began to scale the cliffs behind them. In reaching the top of those cliffs they reached the summit of bravery and planted the flag of tradition. The landing and the subsequent campaign for ever showed the world the mettle of the Australian soldier. Yet it is doubtful if, on that bright and bloody morning, any single one of the Australians engaged gave a thought to posterity or contemplated the effects of his actions, and those of his comrades, on the pattern of Australian life.

How they held the cliffs and gullies, and even in places pushed the Turks back, is beyond comprehension, for the Turk was grim and stubborn in defence and his supply lines were easily kept organised.

The 3rd Brigade were to make the first landing and the 1500 men of the first wave were brought from Imbros in three old battleships. The men and officers of the battleships, for that night, gave up their sleeping quarters to the Australians carried in the ships, insisted on feeding them and providing extras from the ships' canteens, and woke the troops with hot cocoa. These actions and gestures led to intense affection between the Royal Navy and the AIF. Many a British naval officer afterwards remarked how orderly and silent, now that the test had come, were these troops whose reputed doings had shocked Cairo.

The navy was impressed at the way the Australians landed. The Australians, in turn, were surprised at the cool bravery of the young midshipmen and the crews of the whaleboats and other small boats that brought them in to the beaches.

Officers and men tumbled out of the boats and waded ashore. They found themselves facing a country quite different from that described to them in lectures on their tasks. They had been told to rush across the beach and shelter under a bank. They were to drop their packs there, quickly form up, fix bayonets, load their magazines, then advance over a belt of open land to a comparatively low ridge. Then they were supposed to reorganise and push off again towards specified points on a long ridge about 1½ kilometres inland. But the country was nothing like the preliminary description.

There were no signs of life on Gallipoli until the first men were climbing out and were in about a metre of water. The first man ashore is believed to have been Lieutenant D. Chapman of Maryborough, Queensland, of the 9th Battalion, killed at Pozières in 1916. Men in the boats began to be hit but this danger was quickly over when they ran across the shingle and sheltered under the bank. The beach was not the inferno of bursting shells, barbed-wire entanglements and falling men as it has sometimes been painted or described. Nevertheless it was not long before the landing was hotly contested and many men fell.

One boat full of dead men drifted aimlessly in the surf, and bodies caught on the underwater wire laid by the Turks bobbed up and down. Equipment lay everywhere. The sand and the surf seemed to be alive with countless little spit-spats of sand as bullets hissed down from the heights.

Men died trying to drag wounded mates to safety. Some bodies lying on the beach were later found to have as many as twenty bullet wounds. The noise was thunderous. Mingled with the Turk rifle and artillery fire was the heavy barrage from the British naval force covering the landing.

It was at this time that the Diggers' ironic sense of humour first appeared. There is the story that one Digger, on landing, said: 'Hey, they're carrying this joke too far. They're using ball ammunition!'

Singing and shouting, the Anzacs went up the cliffs. Hundreds died on the cliffs, some falling back to the beach, some staying lifeless against the cliff face. But by noon 10,000 men were ashore at Anzac Cove, and some of them were dug in at the top of the cliffs. Fighting went on, without a minute's break, from the moment of landing for nearly forty-eight hours.

Anzac losses were great, but the men hung on to the land they had so dearly won, and they stayed put until ordered out. And that is largely why Australians have ever since been so difficult to dislodge. At Gallipoli they established a reputation for taking and holding against great odds and ever since, perhaps subconsciously, they have taken and held.

After some hours it was apparent that all the carefully laid plans for the invasion were useless, but it was noticeable at every interval in the attack that the officers and NCOs, as they had learnt in their desert training, reorganised the troops whom they found near them, whether or not they belonged to their own units. In spite of the un-nerving confusion which could easily have wrecked the whole undertaking, frequent and clear reports from some of the scattered and furiously fighting parties ahead quickly reached the commanders. Many of these reports came from privates who had taken over the command of sections and even platoons. Because of these reports, senior commanders never at any time entirely lost control of the situation in general. At eight o'clock a few men of the 10th Battalion reached Scrubby Knoll on the third ridge, about 1½ kilometres north-west of the 400 plateau, the farthest point reached during the whole campaign. First to reach it were two scouts, Lance-Corporal P. Robin (who was killed three days later) and Private A. S. Blackburn (who won the VC later in the war and in the Second World War commanded as brigadier the Australian troops left behind and captured in Java). Very early, that first day, it was clear that the rigorous desert training had been worthwhile. Men were firing accurately, directed by orders passed verbally along the firing line, and they had not touched their water-bottles, a sign of high self-discipline.

For a time, on the first day, Captain J. P. Lalor, grandson of Peter Lalor, leader of the Eureka revolt riot in 1854, was senior officer of the 12th Battalion in a certain area. Fiery and impetuous, he wanted to advance but he knew that the 12th was brigade reserve, so while a fellow officer pushed forward with another company Lalor remained, digging a semicircular line of rifle pits just short of the spot known as the Nek. Lalor carried with him a treasured family sword, the sword used by his grandfather at Eureka. Captain Lalor, fighting furiously, was killed during the afternoon and his sword was lost. It was picked up by Lance-Corporal Harry Freem but was dropped again and is said today to be in a Turkish museum, though I have been unable to verify this.

Australian troops have always been anxious to save their wounded and on the very first day of the Gallipoli campaign there were many heroic instances of attempts to rescue wounded. Of 140 officers and men of the 7th Battalion who came ashore on one beach, only thirty-five managed to cross the beach and shelter behind some grassy hummocks. Wounded Australians left in boats were rescued under fire and carried to safety.

Many bodies, particularly those of men lost far out on Pine Ridge, were not found until 1919 when the Australian Historical Mission visited Gallipoli. The mission found the remains of the troops still there in the little crescent-shaped groups in which they had fought it out, and with their battalion colours still visible on their sleeves.

On Gallipoli, as later in France, many Australians were killed because of poor tactics devised by British leaders. This was the case when Australians, New Zealanders and French troops were used in an ill-planned attempt to capture some stone houses at Krithia. The 2nd Brigade, already reduced at Anzac to 2900 men, lost another 1000 in just one hour's fighting. The attack was ill-timed and the Australians, with the New Zealand brigade and some French on either flank, had far too great a distance to travel over open ground. They advanced very quickly and in open order, but the closest they could reach was 370 metres from the Turkish earthworks. Three weeks later an English division, advancing by night, approached another 180 metres nearer the Turkish position almost without casualties.

All too often during World War I Australian troops took

positions only to be forced out later because they did not have the ammunition to defend them. This happened at Quinn's Post on the night of 9 May 1915. Three parties of the 15th Battalion, totalling 100 men, took the trenches opposite them in some furious fighting. Communication trenches were dug from Quinn's to the new positions but the Australians, out of bombs, were in the morning subjected to heavy flank attacks, losing ten officers killed and 200 casualties among the attacking parties and their supports. The Quinn's garrison then consisted of Light Horsemen who had just been brought to Anzac without their horses. They learned from the infantry to catch the Turkish bombs before they burst and throw them back, or to smother them with an overcoat or sandbag. Also, for the first time, bombs were improvised from jam tins filled with snippets of metal, now being manufactured on Anzac beach and used in place of the Mills grenades and other more deadly bombs then coming into use on the Western Front.

The first Australian VC of the Great War was awarded to Private Albert Jacka for an incident which occurred on the morning of 19 May 1915, at Courtney's Post, south of the more famous Quinn's Post, during a determined Turkish attack.

The Turks cleared one bay in a trench system and nine enemy took the post and held it against several costly attempts to clear them. Jacka asked his mates to attract the attention of the Turks. Then, coolly and cleverly taking the calculated risks for which he became famous, Jacka got into the bay held by the Turks, killed six and wounded and captured one.

It was a one-man feat not only of bravery, but of clear thinking, and his mates knew then that Jacka would go far. A tough, clever fighter and leader, Jacka distinguished himself again later in the war. He was often called the bravest man in the AIF and certainly became one of the best known.

Australians have always been vocal in action and many of their war-cries must have sounded odd to foreign ears. On Gallipoli, in answer to the Turks' 'Allah! Allah!' the Anzacs shouted 'Saida baksheesh!' or 'Eggsacook!'. 'Eggsacook!' can sound savage when shouted by a man with his blood up and rifle and bayonet in his hands.

Many Anzacs were killed and wounded through exposing themselves after daylight. This happened not only on Gallipoli but in France. Australians were renowned for the way they walked about on tops of trenches. This was not merely bravado, but reaction at being 'cooped up' in the trenches. The Australian has always respected a game enemy. For this reason he respected and even liked the Turk and the German though this did not prevent him from wanting to kill them.

The Diggers were now battle-hard and battle-wise and like all front-line troops they had developed a resignation, a fatalism. Bean says they felt themselves to be between two long walls from which there seemed to be no turning except death or disabling wounds and this was true enough, for the chances of anybody surviving say four or six attacks was negligible. Many men of the 1st Division gave up hope of ever going home and forgot completely the ambitions of civil life. As the hour for any offensive approached there came over most Australian troops, even over young infantry officers who knew their chances of surviving three or four battles was almost nil, a keenness to make another strike. This meant that the AIF was developing or had by then developed an *esprit de corps*.

Perhaps because they treasure this *esprit de corps* so greatly, Australian 'originals' have always resented reinforcements, who, they suspect, might do something to damage that spirit. This was the case even on Gallipoli, where reinforcements were needed badly enough.

Originals have always disliked, or professed to dislike, reinforcements, who, in their turn became veterans and themselves hand out the same unkind treatment to later drafts. It is a way of showing off, or emphasising the fact that one *is* an original, of impressing the newcomer. Many of the men who died on Gallipoli were reinforcements, which proves that a man does not need to be an original to stop a bullet.

Describing his arrival, as a reinforcement, at Gallipoli, one soldier quoted by E. J. Rule in his book, *Jacka's Mob*, said:

On the side of the ravine nearest the front line an accumulation of dugouts dotted indiscriminately about the slope proved to be the home of our battalion. Later on the inmates began to dribble out in twos and threes, and, coming up to us, stood aloof with a critical look in their eyes. We were not unconscious of a certain pride in our clean clothes, shaven faces and soldierly get-up. They, although they seemed to us more like sewer workers than soldiers, were proud of their rags, their whiskers and the colour of the dirt which stained their clothes and helped to set off their grimy, hairy faces.

As 6 August, the date of the big new offensive against the Turks, approached, the Anzac sick parades diminished. Men already evacuated tried to desert back from ship, hospital or base.

The Digger has always been keen to get out of work and sometimes he will go to extreme lengths to do so. He will also go to extreme lengths to get back into action with his mates. The worst thing that can happen to an Australian infantryman is to have to go into an attack with strangers. He likes to be with his friends.

The Turks earned the respect of the Anzacs. In one attack by 42,000 Turks more than 10,000 were hit and of these 3000 were killed. One reaction to this slaughter, mainly the work of Australian riflemen, was that the attitude of the troops towards the Turks entirely changed. From being bitter and suspicious they became admirers and almost friends with the Turks—'Jacko' or 'Abdul' as they called them—and so they remained until the end of the war.

An attempt by Australians on 20 May to rescue some Turkish wounded led, first, to a short informal truce and later, on 24 May, to a formal truce lasting nearly all day in which the dead were buried.

The Turks took few prisoners at Gallipoli and this made them all the more dangerous. They killed the wounded without compunction and left them to rot. It was at Gallipoli that the Digger found out that human corpses smell, just as animals do. New soldiers, and particularly infantrymen, have been making this discovery for centuries and few have failed to be shocked. On Gallipoli the lesson was learned the hard way, for bodies lay in the open for weeks and sometimes months and the stench hung almost visibly over the peninsula.

That is what any infantryman remembers most about a battlefield—the death smell, searing the nostrils and affronting the senses. To the imaginative man it is almost unbearable, because he sees *himself* dead and rotting and swelling. Men who remain unshaken by shellfire or a bayonet charge can be frightened by the thought of dying alone on the battlefield.

Describing conditions on Gallipoli, E. J. Rule wrote:

We were astonished one morning, when crawling out of our dugouts, to see a mass of snow over everything. Few of the men had seen snow before and, like schoolboys, they ran about pelting each other with snowballs. Later a bitter wind set in from the north, which put an end to frolicking. Clothes which were suitable for the hot weather had to do for the winter, and every man not on duty went to his dugout, rolled himself in his filthy blankets and huddled close to his mates for warmth. Soon the freezing wind turned the snow to ice and foot troubles arose. Many of the men's boots were worn out and some of us had our toes out in the fresh air.

That night our company relieved the front-line garrison. Six waterproof capes were issued to each platoon to be used by the sentries, while the rest of the men sought what shelter they could in the lousy niches cut into the side of the trench. But the howling icy wind, rushing through the bays in the trench made rest impossible. . . . It was not long before those of us with worn-out boots were stumbling about on frozen feet which had lost all feeling. . . . As we were too cold to take our clothes off to chat [delouse] ourselves lice began to irritate our skin. Soon rumours began to float along the line of English sentries found frozen to death at their posts. . . .

The rumours were true. Many men froze to death at their posts, others suffered from trench feet and were taken off to the hospital island of Lemnos to have them amputated. It would have been a bitter winter even with good clothing; it was horrible for the Anzacs in their ragged clothing.

Australian snipers at Gallipoli were helped by an invention of a man of the 2nd Battalion—a periscope rifle. The apparatus was made on the beach and could be safely fired from shelter in the most dangerous posts. This was possibly the first of many hundreds of military inventions by Australians. Using the periscope rifle, the Australian marksman completely routed the Turkish snipers, and Monash Valley, one of the worst danger areas on the whole of Gallipoli, became safe.

Numerous demonstrations were arranged by the Anzac staff to tie Turkish reserves to Anzac and prevent them going elsewhere on the peninsula. These demonstrations ranged from actual attacks and costly sorties against the enemy's parapet, to a sham concentration in which a platoon or two with bayonets showing above the parapets ran like stage soldiers round and round in a circle in the trenches to give the appearance of a battalion assembling to attack. This method was probably the most effective of all and in various ways was subsequently used in France and during the Second World War.

A man had to be very sick indeed before he would consent to being evacuated from Gallipoli. Although gaunt and weak and sometimes even fainting at their posts the men refused to give in even to themselves.

Their uniform [says Bean] was like no other in the war, any degree of undress being sufficient for the men and allowed by their officers.Half naked, they tunnelled, carried food, water and ammunition up the dusty precipitous tracks, swept their trenches free of refuse or patiently searched their clothes for the vermin that nightly plagued them. At no time were troops away from shellfire and there was practically no chance of rest in peaceful conditions. In the gullies where most of them lived immediately behind the front-line they were plagued by Turkish rocket bombs. Anzac beach was a sight perhaps never before seen in modern war—a crowded busy base within half a mile of the centre of the front-line.

"AT THE LANDING, AND HERE EVER SINCE"

Drawn in Blue and Red Pencil by DAVID BARKER

David Barker, an Anzac himself, drew this 'hard case' sketch for *The Anzac Book*, published in 1916.

General Sir William Birdwood, known as 'Birdie', who commanded the Australians and New Zealanders at Gallipoli. (*Imperial War Museum*)

A Digger carries his wounded comrade from the Gallipoli firing line to the medical aid post. (*Imperial War Museum*)

An Australian soldier on sentry duty in a Gallipoli trench while a mate below rests in a scooped-out possie.

Australians of the Imperial Camel Corps near Rafah, Palestine, in January 1918. The cameleers, nearly all of them Light Horse troopers, were used for long-range desert patrols and raids.

Sir John Monash, AIF commander in France, bestows a decoration on a soldier, whose side view shows the voluminous cut of the Australian tunic. (*Imperial War Museum*)

Some of the bravest Diggers were non-combatants—the stretcher-bearers. They might well have had the motto: 'We never refused a call.' There is no cry in battle quite like the one for 'Stretcher-bearers!' It is ominous, desperate and imperative, and for the bearers it spells danger, for there is obviously great danger at the spot where a man has been wounded.

The stretcher-bearers of Gallipoli, by their remarkable gallantry and devotion to duty, set an imposing tradition for their successors. Most famous Gallipoli bearer was Private Simpson (full civil name John Simpson Kirkpatrick), a bearer with the 3rd Field Ambulance. Simpson began his work on Anzac night, when he annexed a donkey used for transport. From that moment, working every day and half of every night, Simpson laboured continuously bringing men from the front to safety on the beaches. He put a Red Cross brassard around his donkey's forehead and the pair became a common sight on Gallipoli.

His colonel, recognising the value of Simpson's work, allowed him to work as a one-man unit. He slept at an Indian mule camp and reported once a day to his own unit. He acquired a second donkey so as to bring even more wounded back. Simpson escaped death so many times that he became completely fatalistic and no amount of fire stopped him from work. A quiet, unassuming man with a gentle nature, Simpson was a Gallipoli celebrity without knowing it. He breakfasted most mornings with the men on water guard at a certain point on the way to the front-line trenches. One morning on his way through he found that breakfast wasn't ready. 'Well, never mind,' he said, 'I won't wait. Get me a good dinner when I come back.' But he never did come back. Enemy fire got him that morning.

Another bearer, who was seen to pass without stopping under a shrapnel burst, carried his patient to the dressing-tent, sat quietly in the waiting ranks for the doctor to come to him and fell dead without having mentioned his own wound.

The beach parties and the naval men (who also helped with the landing of men and stores) carried on under almost any amount of fire. Even the defaulters sent on the open decks of the water-barges to pump water into the tanks on the beach were too proud to turn their heads when shells burst over them. Part of the tradition set at the landing was to carry on under shellfire as heedless as if the shrapnel was a summer shower, and this applied from Birdwood down to the latest reinforcement.

Bomb-fighting in the battle of 6 August was fiercer than the Australians had ever seen—even at Quinn's Post. Finding that the Turkish bombs had long fuses the Australians constantly caught them and threw them back before they burst. Unfortunately, the Turks themselves were not slow to learn; they shortened the fuses and many Australians had their hands blown off and others were blinded or killed. In World War II the throwing back of bombs was strictly forbidden and I know of no instance where it was practised.

Fighting at Lone Pine ceased on 10 August but by then six Australian battalions had lost in all eighty officers and 2197 men. The 3rd Battalion had every officer hit except the quartermaster. Battalions of the 1st Brigade lost very heavily and few witnesses of its service remained. Consequently, of the seven Victoria Crosses awarded after this fight, four went to a reinforcing battalion, the 7th. It was only small consolation that the Turks lost 5000 men.

The most brilliant operation of the entire offensive was carried out by New Zealanders on Bauchop's Hill. The New Zealanders, strictly obeying the order that only bayonets were to be used and that there must be no cheers, dashed suddenly and silently through the dark from one Turkish post to the next, coming on them from unexpected directions. They even wrestled with the Turks in some cases.

Australian tunnellers first became active on Gallipoli and at Lone Pine and the Nek several times broke into Turkish networks. Along at least one of these tunnels at Lone Pine the miners crawled close enough to see the legs of the Turkish garrison. The Turkish miners were everywhere outdug and outfought and by December 1915 several Anzac tunnels were under or near the Turkish front-line, ready for explosion.

What hurt the Australians and New Zealanders most about leaving Gallipoli was having to abandon the bodies of the mates who had been killed. From 12 December onwards the cemeteries of Anzac were never without men tidying up the graves of dead mates and repairing or renewing the little packing-wood crosses and rough inscriptions.

The masterpiece of the evacuation plan was drawn up by General Brudenell White—in a paper which has been described as a model of precision and clear thinking—and carried out under his close supervision.

Many Anzacs felt they were 'running away like dingoes'. Originals have told me that they regarded it as the most shameful day of their lives. Later, they realised that evacuation was inevitable and began to feel a great pride in the way that the Turks had been deceived into believing that the British force was still in possession while in fact it was being evacuated without a single casualty.

It was during the evacuation that the Digger's great ability to improvise came into its own. The Turk had to be tricked. The Australians were just the ones to do it. While the main force was taken down to the beach and evacuated at night, the men of the small rear-guard had to keep the great bluff going for twenty-four hours. Their duty was to move up and down the trenches firing off their rifles to keep up the normal sound, and, during the following day, to give the impression of a much larger force by being here, there and everywhere.

But they did more than that. To ensure that rifles would fire even after the last man had left, some of the Australians fixed leaking tins of water above empty tins on to the triggers of rifles below. When the weight became sufficient the rifle would fire. Some tins were set to leak fast, others slow. A great variety of booby-traps were left for the Turks when eventually they came down to investigate.

In Ankara in 1957, I spoke to one of the first Turks to probe the defences after the Australians had gone. Through an interpreter he told me, 'Our unit was for some time puzzled by the lack of movement in the British lines. Usually we could see men moving about during the day, or spot smoke from cooking fires. But now there was nothing, even though rifles continued to fire, and even machine-guns. Much more fire came from the Australian lines than from elsewhere, but it was still less than we were accustomed to.

'Night patrols met no real opposition and that surprised us, too, but we decided the British tactics had changed. I led a patrol into the Australian lines and the silence was so uncanny it frightened us. Every now and then a rifle would fire, but not directly at us, though some Turks were killed and wounded by this phantom firing.

'When, eventually, we raided the Australian trenches and found them deserted we were astounded. Even then we thought it was some kind of trap. I, personally, thought the whole area had been mined and that somewhere the Australians were hidden waiting to blow us up the moment they felt they had enough victims in the bag.

'It was a completely successful deception and it must have saved thousands of lives. Had we caught them evacuating we could have killed as many troops as we had killed arriving.'

From the first rumour of the evacuation men demanded 'as a right' to be included in the last parties. 'I was first and I have a right to be last' was a frequent argument. Some men asked to be paraded before their COs to discover what they had done to be denied this 'right'. All sorts of private schemes were hatched in attempts to stay until the last.

For the withdrawing Anzacs Gallipoli had a special meaning. It was not merely that 7600 Australians and nearly 2500 New Zealanders had been killed or mortally wounded there and 24,000 more (19,000 Australians and 5000 New Zealanders) had been wounded, while less than 100 were prisoners. The standards set by the first companies, by the stretcher-bearers, the medical officers, the staffs, the company leaders, the privates, the defaulters on the water-barges, the Light Horse at the Nek, were already part of the tradition not only of Anzac but the Australian and New Zealand people. Before Gallipoli was abandoned the word Anzac stood for 'reckless valour and a good cause, for enterprise, resourcefulness, fidelity, comradeship and endurance' (Bean).

The Gallipoli campaign was a failure and most countries would have been glad to forget such an obvious débâcle.

Australia was not content to forget. The defeat became a victory and Anzac became a symbol of something greater than most Australians realise. It was important in the evolution of the Australian soldier. No action before that had so strongly tested him; and probably no action since has been so severe, though the trenches of the Western Front were quite bad enough. Conditions in Tobruk were not good, but by Gallipoli standards they were not bad; the Kokoda Trail was really bad, but battle strain was not incessant. Gallipoli lasted eight months and the strain and discomfort were unremitting.

The Digger, born in South Africa, grew up on Gallipoli. He returned to Egypt with an outstanding reputation, with a great deal of sober confidence, but most important of all, he had established the Digger character, he had created a type that has ever since been faithfully followed.

Still, it is wrong to suppose—as many writers have supposed—that Digger mateship was born on Gallipoli. Mateship began in South Africa, when the Australians surprised the English troops by the extraordinary way in which they stood by one another, in action and out of it. This mateship was carried over into the AIF by old soldiers of the Boer War, and it turned up again on Gallipoli.

The Australian Goverment did not strike a Gallipoli medal. Such a medal was thought of, but somebody in the Australian Government thought that as the British Government had not awarded a medal or clasp to British troops the issue of an Australian medal might give offence.

The Anzacs did not seek a medal for Gallipoli and were content with the brass A which they were allowed to mount on their colour patches, but it would have been an historic gesture to give them a medal and it would have helped to perpetuate the Anzac tradition. The medal was actually designed and the ribbon made. The medal was to be called either the Anzac or the Gallipoli Star.

After many years of pressure the Australian Government in 1967 issued the Anzac Commemorative Medallion to surviving members of the AIF who had served on the Gallipoli Peninsula at any time between 25 April and the date of final evacuation. The eloquent medallion, designed by Raymond Ewers, shows Simpson on his donkey on the obverse. By 1967 about 10,000 Anzac veterans were still living to claim their medallions; it was a shabby way to keep down the cost.

6

Diggers in Red Capes

Nurses enlisted to go to the war for much the same reasons as the soldiers; they were eager for adventure and foreign travel and they wanted 'to be in it'. My mother told me that she knew 'the boys would need help'. She also said, 'There was the prospect of travel and I had always wanted to do that. And I was interested in nursing. I expected to learn more at the war, and, my God, I did.' She knew nurses who enlisted because they had brothers or sweethearts in the AIF.

Most of the nurses joined the Australian Army Nursing Service but some women volunteered for the Queen Alexandra Imperial Military Nursing Service Reserve, whose members worked in British Army hospitals. These girls were in a hurry to get to the war and they thought that QAIMNSR gave them a better chance than AANS. In some cases, they were right; they were dealing with wounds and battle injuries in France before the AANS girls received their first Gallipoli casualties.

The AANS sisters were given 'terms of engagement' which entitled them to passage from and to Australia 'by 2nd class mail steamer' or by military transport. The earliest nurses travelled on the troopship *Benalla* in the first convoy.

Sisters were paid £50 a year and staff nurses £40, with an allowance of up to 21 shillings a week. There was also an allowance of £10.10 for their uniform, an ankle-length grey one with red cape and white veil, an outfit which looked both attractive and military. To enhance their authority they held honorary commissioned rank. By July 1915 they were being paid by daily rate, as the soldiers were. A sister received 9s 6d and a staff nurse 7s.

Nurses got no closer to Gallipoli than the island of Lemnos. My mother, who served there, described the island and the work of the nurses:

When Byron raved about 'The Isles of Greece; the Isles of Greece!' he obviously was not referring to Lemnos Island. It is a barren, stony land with here and there as oasis. Hot, very hot in summer, bitterly cold in winter. The villagers had nothing to give and nothing to sell. The English, Canadians and Australians had set up hospitals, under canvas, to nurse the sick and wounded from Gallipoli; ours was at the village of Mudros on an exposed headland. There was no water; villagers got theirs from a communal pump, ours was brought to us in ships. It was strictly rationed—we had no baths or showers— but we could and did go to the cookhouse each evening for a can of hot water which we took to our tents and bathed in an enamel dish. No laundering of linen could be attempted, so a marquee was requisitioned for the storage of soiled linen. The villagers did ours, in a kind of a fashion. Later a condenser was built and salt extracted from sea-water, pipes were laid and taps appeared here and there and water freely flowed. However, it was too late for the bed-linen, and it was sad to see the lovely Red Cross issue of sheets, pillow-cases, towels and pyjamas going up in smoke.

Army rations were crude, so were our cooks. They were trying to feed about 1500 people, under extremely adverse conditions. We kept ourselves alive by the Red Cross issue of Ideal Milk, tins of coffee *au lait*, Huntley and Palmer biscuits. We, in turn, kept our patients alive by the same means plus soup made from dried cubes and cooked over an open outdoor fire in a dixie, bully beef, army biscuits, salty bacon, badly cooked porridge, prunes, rice and straw. Straw was not an issue, it just got in during the cooking process. We had no bread, no butter, no margarine.

Later bakers' ovens were made of large drainage pipes, but the bread which issued forth was blue in colour and uneatable for most of us. We preferred to split the army biscuits in layers and spread them with jam or soak them in tea to soften them.

Just before we left Lemnos canteens began to appear and we could buy tinned fruit and chocolate bars.

I did get a piece of steak one night. While the blizzard raged and dinner-time came round—7 p.m.—Silas, my orderly, said: 'Don't go down to the mess tonight. I'll scrounge round and see what I can find. No good going down there. What will they give you, even supposing you even get there in this wind?' He went away and came back with a large piece of steak. It seemd there was good food on the ships in the harbour! It was the first fresh meat I'd seen for several months. Silas cooked it and we also had a tin of peaches and made coffee.

One morning I was at a church service, conducted in the open air by Dean Talbot of Sydney. There were hundreds of soldiers and a few nurses and we all stood, of course. After it was over and we were about to disperse, a tall, bronzed and bearded soldier—he was twenty-one years old—approached and put his arms round me, saying as he kissed me, 'Little Sister.' It was Bill Rose who had been at the Gallipoli landing and whom I had last seen in Sydney. He explained that his battalion had been sent to Garpi Rest Camp, two miles distant, for a rest. It was now September and they had been on the peninsula since 25 April.

'We only arrived last night, but I had to come and look for you,' he said.

We talked for a while and then he said he would go back to the camp for a clean-up and would be over again at 7 p.m. when I would be free.

He would bring two mates, he said. They were Rosser and Arthur Hynes. Rosser was later killed. I had two nurses with me and we had to be very circumspect. We seemingly parted, but met a few minutes later, beneath the brow of the hill, unseen by those in the camp above. The men wore shorts and they flicked mosquitoes from their bare legs, as we talked and listened to the booming of the guns on Gallipoli, twenty-five miles away. Bill explained how they used the bayonet. 'Plunge it in,' he said, 'give it a twist, to let the air in. That's what that groove is for.' He pointed to the indentation, which runs along the blade. [This was a popular misconception.]

They were young and keen and we were shocked, of course.

They came every night after that and they would be waiting for us as we came off duty. The party broke up when I was put on night duty. But Bill was able to get letters across. Then one night, one of my orderlies came to tell me there was a sergeant outside to see me. It was Bill. He said they had sudden orders to return to Gallipoli. 'I have broken camp but I had to see you to say good-bye,' he said.

My presence outside my ward tent with this soldier was strictly against the rules, but the orderly said: 'Don't worry, Sis, I'll keep a good watch, you go into the day tent and say good-bye properly.'

Next morning as the night staff were going off duty we heard bands playing and presently men marching four abreast came down the hill, on their way to the ships which would take them back to the slaughter. They were in high spirits. Bill saw me there, had a word to his OC, left the ranks and took me by the hand. 'Walk with me a little way,' he said. So hand-in-hand we walked until his section was nearly out of sight. Once again there was a brief farewell and as he embraced me, the following troops were much amused and a band struck up 'Good-bye Dolly, I must leave you'.

There was no time to be sad for long with one's personal sorrow when men were dying with dysentery in the wards. Three marquees connected by canvas hallways constituted a ward. It was heartbreaking to see their wasted frames and sunken eyes—every ounce of fluid drained from their bodies. Next disaster was the blizzard. We were plunged from great heat into extreme cold, with thunder, wind and rain more violent than any man of our army had ever seen. Our harvest was frost-bitten feet. There was extreme suffering. Most of those afflicted were from Suvla and therefore not Australians but English. They were little men, almost dwarfs, from the hills and the mines of Lancashire and Yorkshire. They were conscripts and known as Kitchener's Army, that is, they had no pre-war military training.

Their suffering was so intense they cried as little children do, loudly. Most, I think, recovered the use of their feet, but some lost toes, half a foot, a whole foot or maybe both feet.

Amidst the suffering and the crying, there was humour. I remember an attempt by one man to lead the others in song. He could win no followers, so he kept on alone, doggedly singing over and over again—'Though your heart may ache awhile never mind, there'll be sunshine after rain. There'll be gladness after pain—never mind.'

Huts were in the course of erection after the blizzard, in a less exposed position, it being considered that neither we nor our patients would survive the winter under canvas. We had no suitable clothing, so were issued with soldiers' greatcoats, boots, puttees and balaclavas. Our own footwear came off in the mud.

There was great calm after the blizzard and a pale weak sun shone. Walking was our only pastime. My friend, Sister R. Dickenson, and I went to the British cemetery, to see the graves of two Canadian sisters who had died of dysentery. [Sister Dickenson died during the influenza epidemic of 1918 and is buried in the military section of the churchyard at Harefield, Middlesex.] We were astounded to see rows and rows of open graves, each with its mound of earth, the exact counterpart of each other. We asked the soldier grave-diggers what they were for, but they knew no more than we did, they were obeying orders. We knew our patients were not dying in such vast numbers. It was not until after the evacuation of Gallipoli that the reason became clear. The graves were, of course, for dead expected during the evacuation. What satisfaction it must have been, to shovel back the earth, minus a soldier in a grey blanket. There were no coffins on Lemnos.

The next excitement was the evacuation of Gallipoli. I was on night duty again, at 3 a.m. working away with my three orderlies among the sixty-odd patients. One orderly came up and said: 'Something is going on down at the embarkation wharf. There is a rumour that we have left Gallipoli.'

My reply was: 'How absurd. How can we leave Gallipoli?'

'If you can spare me, I'll scout round and see what I can find out,' he said. He came back with the news that there seemed to be hundreds of ships crammed with soldiers and true enough they had left Gallipoli and even now were coming ashore. As each ship unloaded its human freight and moved away another pulled in and did likewise. My marquees were near the roadside and soon we could hear the tramp of marching feet.

At 7 a.m. we were once again coming off duty and watching men, clothes caked with mud, bowed down with their equipment and souvenirs, marching up the hill, down which they had come so blithely a few short weeks before. There was no semblance of order; just men, of every British race, colour and creed. Thousands and thousands. All day they marched and it was night before they were all gathered in at Garpi Rest Camp. But where was Bill Rose, where was Rosser, where were all the others I had known? One of my orderlies was a good scout and he found out that Bill had been seen boarding one of the ships and as there were no casualties he must have got away. I met him again in Cairo and had many outings with him. We had a final farewell there the day he went to France, where he was killed. [Company Sergeant-Major William Henry Rose, 55th Battalion; killed in action 20 July 1916 aged twenty-one and commemorated on the wall of VC Corner, Australian Cemetery, Fromelles, France.]

It was a great privilege to be on Lemnos, so close to Gallipoli, nursing wounded and ill soldiers. The Australians were all stoics, never complaining about their wounds or pain—though, as usual, they grouched enough about the army in general. We knew they didn't mean a tenth of their grouches.

Even when they died they made no fuss. Doctors and nurses felt helpless and useless at times. The wounded men often wanted to know what had happened to a particular mate and would Sister please find out? We would make inquiries at the other wards and take back the information. Sometimes men were transferred from one ward to another to be with their mates.

I think that most of them were irritated with themselves for being wounded or sick—as if they had got that way through some fault of their own. They felt they were letting down their comrades 'back there' and sometimes it was very difficult to know if a man was as fit as he claimed to be. Some men, though very ill, perhaps with half-healed wounds, would report fit so that they could get back to their units— where, of course, they would have been more a hindrance than a help. I don't think they put it over the sisters very often.

Gallipoli was a severe test for the Army Medical Corps, but they came through it as well as the soldiers on Gallipoli and the experience doctors and nurses had on Lemnos saved many lives in France during the next three years.

7

'Fierce Joy of Battle';
Fierce Creed of Mateship

In a foreword to a book on the Great War, John Masefield wrote a considered assessment of the Australian soldier from his own observation and experience of the Diggers over a three-year period:

During the war the English suddenly became aware of a new kind of man, unlike any usually seen here. These strangers were not Europeans; they were not Americans. They seemed to be of the one race, for all of them had something of the same bearing, and something of the same look of humorous, swift decision. On the whole they were taller, broader, better-looking and more graceful in their movements than other races.

Yet in spite of so much power and beauty they were very friendly people, easy to get on with, most helpful, kind and hospitable. Though they were all in uniform, like the rest of Europe, they were remarkable in that their uniform was based upon sense, not upon nonsense. Instead of an idiotic cap that provided no shade for the eyes no screen for the back of the neck, that would not stay on in a wind, nor help to disguise the wearer from air observation, these men wore comfortable soft felt slouch hats, that protected in all weathers and at all times looked well. Instead of idiotic clothes designed for appearance on a parade ground, these men wore clothes in which they could do the hardest of work and then fight for their lives. Instead of bright buttons and badges, 'without which' a general once said to me, 'no discipline could be maintained', these men carried in their equipment nothing that added to the worries of war. When people asked, who are these fellows, nobody, at first, knew.

The strangers became conspicuous in England after about a year of war. They were preceded by the legend that they had been 'difficult' in Egypt, and that they had to be camped in the desert to keep them from throwing Cairo down the Nile. Then came stories of their extraordinary prowess in war. Not even the vigilance of the censors could keep down the accounts of their glory in battle. For themselves, they were a very modest company, whom sometimes one could hear singing to the tune of 'The Church's One Foundation':

We are the Anzac Army,
The A.N.Z.A.C.,
We cannot shoot, we don't salute,
What bloody good are we?
And when we get to Ber-lin
The Kaiser he will say,
'Hoch, hoch! Mein Gott, what a bloody odd lot
To get six bob a day!'

Since that time, the Australian Army has become famous all over the world as the finest army engaged in the Great War. They did not always salute; they did not see the use of it; they did, from time to time, fling parts of Cairo down the Nile and some of them kept the military police alert in most of the back areas. But in battle they were superb. When the Australians were put in, a desperate feat was expected and then done. Every great battle in the West was an honour and more upon their banners.

No such body of free men has given so heroically since our history began.

This was high praise from an Englishman—and from a writer not usually given to extravagant praise. But Masefield was writing from spontaneous enthusiasm for the Diggers, whom he understood.

His opinion found support in *The Times History of the War.* *The Times*, as befitted its serious character, was not so generously complimentary, for it noted that the Australians had their shortcomings. But *The Times* study, in general, was the same as Masefield's:

> Special recognition is due to the troops from the selfgov-
> erning Dominions who gave us their best without stint
> and for whom this war has marked their coming into full
> nationhood. United as they are in the common devotion
> to the Allied cause there were still well marked differences
> between the character of the Dominion troops and their
> military methods.
>
> The Canadian troops are distinguished by a seriousness
> in the field which contrasts both with the apparent non-
> chalance of the British private soldier and the ardour of
> the Australians. They (the Canadians) did nothing in the
> war that was not good, and, as their numbers and their
> experience grew, they became, with the Australian *corps
> d'élite*, always sure to be there when there was particularly
> difficult work afoot. In the last three years of the war
> their numbers were kept up to a quarter of a million men,
> and their casualties from first to last—nearly 60,000 officers
> and men killed and nearly 150,000 wounded—show the
> character of the fighting in which they were engaged.
>
> Compared with the Canadians the Australians had less
> steadiness and phlegm, but more brilliancy and dash. The
> Canadian took his fighting like the Englishman, without
> any display of enthusiasm, and perhaps, on that very ac-
> count was the more dogged and stubborn in defeat. The
> Australians undoubtedly had more of the fierce joy of battle
> than any of the troops engaged on the Western Front and
> Macauley's celebrated description of Cromwell's Ironsides
> has not been inaptly applied to them by one of their recent
> historians, Mr Cutlack [F. M. Cutlack, author of *The Aus-
> tralians: Their Final Campaign, 1918*].
>
> *They moved to victory with the precision of machines while
> burning with the wildest fanaticism of the Crusaders. . . . They
> marched against the most renowned battalions of Europe with
> disdainful confidence. Turenne was startled by the shout of ex-
> ultation with which his English allies advanced to the combat
> and expressed the delight of a true soldier when he learned that
> it was ever the fashion of Cromwell's pikemen to rejoice greatly
> when they saw the enemy.*
>
> It would hardly be true to credit the Australians with
> the 'rigid discipline' of the Ironsides, but their indiscipline

was more in the form than in the essentials of military life and it expressed the free democratic spirit of their country, not any indifference to the value of obedience and co-operation in the moment of danger. Nor was their zeal for the fight mere recklessness. On the contrary, they fought cleverly as well as valiantly, had the keenest respect for the intellectual side of war, a good eye for ground and rare instinct for all the craftsmanship of their new job. In spite of their tendency to go out of hand when they had no serious work they were liked as well as admired in France.

It struck the imagination of the French people that these men from the other side of the world should be fighting the battle of freedom on their soil. M. Clemenceau in a speech he made to the Australians just before the beginning of the Allied offensive showed a keen appreciation of the political romance of their appearance on the battlefields of France.

'We have all been fighting,' he said, 'the same battle of freedom on these old battlegrounds. You have all heard the names of them in history. But it is a great wonder, too, in history, that you should be here fighting on the old battlefields which you never thought, perhaps, to see. . . . We knew you would fight a real fight, but we did not know that from the very beginning you would astonish the whole continent with your valour.'

His praise was richly deserved.

It is misleading to say that the Diggers moved with the precision of machines; anything less machine-like than the Digger is hard to imagine, for the term implies a robot-like being only capable of rigid adherence to instructions and devoid of initiative and individual enterprise, when, in fact, these two qualities are the most marked of all Digger traits.

It is also wrong to credit (or debit?) them with any wild fanaticism. The Digger is a fatalist rather than a fanatic. In any case, it is impossible for a cynic to be a fanatic and the Digger is an arrant and blatant cynic.

I think it is true to say that the Digger has always rejoiced to see the enemy, for this meant, if nothing else, that there was some point after all, to his being in a trench or foxhole.

The sight of the enemy meant action and purpose. He exulted not because a blood-bath was imminent but because he had enlisted to fight and fighting had at last arrived. (Many Diggers of World War II hated the Japanese, but I can recall only two men in my experience who really hated the enemy with a kind of religious fervour. They exulted in killing and both were themselves killed because their very hatred led them to attempt foolhardy tasks in battle.)

At the end of the Great War, the Allied leader, Marshal Foch, said: 'From start to finish the Australians distinguished themselves by their endurance and boldness. By their initiative, their fighting spirit, their magnificent ardour, they proved themselves to be shock troops of the first order.'

Despite such unsolicited opinions, the Australian public could not believe that the AIF was as valiant and remarkable as it appeared to be. What had these men done to merit such comments from famous men?

The Gallipoli campaign and the war's unfavourable progress caused heavy enlistments in Australia. This was neither the first nor last time that Australians responded to adversity; enlistments have always been heaviest when the future has been blackest.

Early in 1916 the Anzac forces in Egypt were reorganised by Generals Birdwood and White into double the previous number of divisions—four Australian and one New Zealand. An extra Australian division was being raised and sent to England and a mixed Anzac Mounted Division was also formed. With the largely Australian Camel Corps it stayed in Egypt and Sinai, where it continued to add to the Digger legend.

When the reorganisation was finished the 1st Anzac Corps, comprising the old divisions, sailed for France, leaving the new 2nd Anzac Corps to continue their training.

When the 1st Anzac Corps reached Marseilles on 19 March 1916, the British staff was prepared for an expected emergency—that the Australians might break out in a riot into Marseilles. But partly through the fine control of the Australian officers the fear was groundless. The Marseilles authorities afterwards told Birdwood's staff that no troops had ever given less trouble. Later, when the 5th Australian Division passed through the city the British commandant at Marseilles wrote to its commander that 'not a single case of misbehaviour or lack of discipline has been brought to my notice. This is a record'.

The Diggers were too interested, too eager to reach the Western Front to give any trouble. The three months between May and July 1916 were, for most Australian soldiers, their outstanding experience of trench warfare in its most settled form. About half the troops had learned trench warfare in Gallipoli but here in Flanders it was, to start with, less tense and more comfortable. There were shops in the villages and *estaminets* selling Flemish beer and cheap wine. At least one *estaminet* was only about 700 metres from the firing line. Cold weather and the presence of women, plus the efforts of the staff officers, resulted in the disappearance of the 'Anzac uniform' and an increased tidiness in dress. It seldom rivalled the tight smartness of the British uniforms but nevertheless it became noted as the most workman-like uniform on the whole of the Western Front.

Despite training in Egypt no regulations or explanations could induce the Australians to salute all officers met, at least when off duty. But they almost invariably saluted officers of their own battalions, men for whom they had the most profound respect—respect bought the hard way, in action. The morale of the troops at this time was particularly high, partly in consequence of a story that leave to England was to be allowed at yearly intervals. England was only 113 kilometres away and nearly all the Australians wanted to go there. Many troops later arriving in the theatre of war went to England before proceeding to France.

During these three months Australians were equipped with the British-type steel helmets and small box-respirators. They disliked both but on the Western Front they saved many lives. For a time the Diggers regarded the Western Front as a picnic. Sniping was harmless after Gallipoli and there was nothing like the tension that existed at Anzac. But already there were signs to the discerning that in many ways this was likely to be a tougher campaign than at Anzac.

Australians in that war were careless in exposing themselves to view. They came, by many bitter experiences, to
realise at last that their movements were patiently watched
and methodically noted by the thorough Germans. Spies did
exist in the towns behind the line and the current talk of
the back areas of each side normally reached the other side
by a neutral country in a week, yet by far the most abundant
and important information secured by each side was obtained
by its forward troops patiently watching their opponent and
possibly capturing a soldier for close questioning. In the
nightly patrolling of no-man's-land, while the Diggers prided
themselves on clever and vigorous scouting, the Germans at
this time were well-trained, bold and persistent.

On the night of 5 May the Germans heavily shelled part
of the Australian line near the Bridioux salient. Two hours
later when the gunfire lifted the trenches were a shambles,
a mass of tumbled sandbags and other debris. Later the
German wireless announced that the Germans had captured
some prisoners, two machine-guns and two trench mortars.
The mortars were Britain's secret weapon. Stokes' mortars,
they should have been brought back on the previous day
after their shoot but an officer had carelessly left them in the
front-line for further use. It was a long time before the
Australian infantry in France lived down the results of this
incident. This was all the more mortifying because some
British newspapers were making copy out of Anzac heroics.
The Diggers themselves were most conscious of being publicised at a time when they were still untried on this front.
The Germans attached no special importance to the mortars
and it is believed that they were never recognised as Stokes'
guns.

The first Australian raid was made on the night of 5 June
on the outskirts of Armentières by a raiding party of the 7th
Brigade led by Captain Maitland Foss, a West Australian
farmer. The raid was carefully practised and each section
knew their particular work. The raid was a great success and
raised the Diggers' confidence. Their own men had been in
the German trenches and their artillery had dealt with the
enemy as it pleased. The raid was quickly followed by several

others, each Australian brigade having prepared one or more.

Twenty-one prisoners were taken in a famous raid by parties of the 9th Battalion led by Captain Wilder-Neligan just north of the Sugar Loaf salient. This raiding made the German infantry's well-tried Bavarian and Prussian troops very nervous indeed.

The creation of artillery for the two new Australian divisions, the 4th and 5th, was a task unparalleled in British experience as the classic example of the speed with which Australians can be trained.

To make up the 3000 artillery officers and men required for each division, 150 trained officers and men from the older divisions were allotted to each. Within a fortnight the new officers were getting results. They themselves were lectured for four hours daily, then they instructed their men for eight hours.

Scouting and patrolling were exciting and challenging, even satisfying to the Australian temperament, but the Diggers' first experience of pitched battle at close quarters and under sustained shellfire brought them face to face with the brutal reality of trench warfare.

It happened at Fromelles, near Armentières, which began on 19 July 1916. It was really part of the great battle of the Somme, which had begun on 1 July. On that first day the British suffered 60,000 casualties, including 20,000 killed. The offensive was still going badly and to stop the German high command from moving troops south to stiffen even further the resistance on the Somme, General Haig and his staff proposed to block the Germans at Fromelles with the Australian 5th Division (Lieut.-General J. W. McCay) and the 61st British Division.

Completely mismanaged by Lieut.-General R. Haking, commander of XI Corps, the battle was a bloody débâcle as the Australian Chief-of-Staff, Brudenell White, had forecast that it would be. The Australians were heavily shelled before they went over the top but they cleared the Germans from their front and support trenches. Their objective was a supposed third trench line but it did not exist; the British Staff had made a mistake. The three Australian brigades involved captured about 900 metres of enemy trenches—a tremendous achievement in such warfare. A British attack that was supposed to coincide with an Australian attack was called off—but nobody told the Australians. They were thus exposed and outflanked.

AIF carrying parties, having delivered their loads of bombs, ammunition and other supplies, stayed in the front line to fight with the men already there. The position became desperate when the Germans reoccupied their old frontline, that is, behind the advanced Australians. In this crisis Captain C. Arblaster, aged twenty-one, led a charge *to the rear*. Arblaster was wounded and died in German hands five days later.

Small parties of Diggers extricated themselves from the shambles; stretcher-bearers by the score were hit as they tried to bring in wounded. In twenty-seven hours the 5th Division lost 5533 officers and men; 400 were taken prisoner.

General Haking said that his battle plan failed solely because the infantry was new to battle. It failed for many reasons—but this factor was not one of them. The battle was a reckless misconception and it left the AIF profoundly mistrustful of British leadership.

In any analysis of Digger characteristics Fromelles is important for it demonstrated mateship in its purest form—self-sacrifice. Many wounded lay in front of the 5th Division's lines, apparently without hope of rescue. Major A. W. Murdoch of the 29th Battalion, disobeying standing orders imposed by Haig himself, fashioned a Red Cross flag and with one soldier crossed no-man's-land, distributing water bottles on the way. At the German wire he arranged a truce with a Bavarian lieutenant so that the wounded could be collected. Murdoch offered himself as a hostage until the collection was finished. It ended prematurely when General McCay, under pressure from Haking, ordered stretcher-bearers not to go out.

Ending of the truce did not end the rescue work. For three days and nights small groups of Australians—sometimes even one man—went out and brought in wounded, despite the great risk under fire. More than 300 men, who would otherwise have certainly perished, were saved. It was a remarkable demonstration of the creed of mateship.

In the 1930s a Digger who took part in the Tel el Kebir to Ferry Post desert march in 1915 told me that it provided, for him, the best example of mateship of the entire war. On the second day the men covered thirty kilometres under blazing sun, through soft sand and without water. After marching from six in the morning until five in the afternoon he saw two exhausted men stagger across the desert for about 1½ kilometres, carrying into camp an unconscious mate between them.

8

'Densely Sewn with Australian Sacrifice'

Fromelles was a prelude to even worse fighting. With little warning and less preparation Major-General H. B. Walker was ordered 'to go into the line and take Pozières' with the AIF 1st Division. Along the ridge just behind the ruined village of Pozières the Germans, strongly dug in, were still holding up the Somme offensive three weeks after it had begun. British infantry had tried four times to take Pozières ridge. Now aware of the Australians' fighting spirit, Haig believed that he had found a cutting edge for another attack on the ridge.

On Sunday, 23 July, the 3rd and 1st Brigades of the 1st Australian Division attacked at Pozières and achieved a striking success at a time when every other portion of the British attack had been repulsed and beaten. While they dug in that night and next morning, the Australian troops, smoking German cigars, were in shiny, black, spiked German helmets, well aware that they had done something spectacular. But soon after this, in a three-day bombardment, the 1st Division lost 5285 officers and men.

The strain was intense. Many officers and men broke down temporarily with intolerable overstrain. In the fighting at Pozières the 2nd Australian Division in a twelve-day tour lost 6848 officers and men. Five of its battalions each lost between 500 and 700 men, almost wiping them out. Even in the 4th Division in the last two days the 48th Battalion lost 600 men, and the 45th Battalion 350.

Against the advice of front-line officers an attack was made on Mouquet Farm. A quarry near the farm was captured and a company under Captain Harry Murray of the 13th Battalion (who was to win a VC and DSO and bar, and an unrivalled reputation as a fighting leader) seized part of the German Fabeck trench, north-east of the farm. Here Murray and his men were outflanked by the Germans who had been ordered by their corps commander not to permit the British plan to develop. But Murray, a former miner, fought his way back with his men in one of the most ably conducted actions in Australian experience.

When the survivors of Pozières came out of the line a man who saw them wrote at the time in his diary: 'The way was absolutely open to shellfire and while others were bending low and running hurriedly our men were walking as if they were in Pitt Street, erect and not hurrying, each man carrying himself as proudly and carelessly as a British officer.' But E. J. Rule wrote:

The fiercest fighting was going on ahead of us. Pozières had been taken by the 1st Division and now the 2nd was going to attack the ridge. When the 2nd Division relieved it, the 1st came back through Warloy. They came by one morning early, having spent the night around Albert and Senlis. Although we knew it was stiff fighting we had our eyes opened when we saw these men march by. They looked like men who had been in hell. Almost without exception each man looked drawn and haggard and so dazed that they appeared to be walking in a dream, and their eyes looked glassy and starey. Quite a few were silly and these were the only noisy ones in the crowd. . . . We noticed that they had lost a lot of men—some companies seemed to have been nearly wiped out, and then again others seemed as if they had not fared quite so badly. In all my experience I had never seen men so shaken up. . . . The shellfire at Pozières was as bad as any during the war.

Describing trench warfare, Rule wrote:

We had not gone far before we found that our trench had been blown in. Dead men were lying along it. Some were partly buried with just an arm or a leg sticking out, and it was here that I came to know what the spongy feeling underfoot meant. It was the first time I had ever scrambled over dead and I was more than terrified; I had never before been so horrified in my life. The shelling, and the dead lying in all sorts of attitudes, were enough to send new men mad.

Pozières was an epic operation by any standards. The work of the 1st Division had made it possible for the 2nd Division to capture the ridge and on the night of 5 August the 4th Division relieved the 2nd. Success brought a dreadful penalty because in the Australian sector there was now a bulge in the German defences. This made the trenches vulnerable to shell-fire from several directions. Also, the alarmed German high command ordered counter-attacks by 'successive waves' of troops, deliberately making them on narrow fronts so as to intensify the shock impact.

On the morning of 7 August the Germans attacked a stretch of defences about 400 metres long and captured some outposts, where they established machine-gun posts. This was where Lieutenant Albert Jacka again showed his courage and leadership. With about eight men who survived from his platoon he took cover in a dugout during a bombardment; one of these men was killed when a German rolled a grenade into the dugout.

Jacka told his men to follow him as he charged into the open to see Germans marching a column of Australian prisoners to the rear, passing through other Germans who were advancing. At that moment the Pozières ridge was in danger of being lost. Jacka and his squad killed the prisoners' escort with their bayonets; the prisoners grabbed the Germans' rifles and Jacka led the whole group in a charge against the Germans who were now in front of the Australians. Clubbing, bayoneting and shooting at pointblank range, they stopped the enemy advance and took prisoner the surviving Germans. In getting back to his own lines Jacka was seriously wounded.

Jacka's actions had alone broken the German attempt to retake the ridge. The 4th Division were so proud of him that they took to calling themselves 'Jacka's Mob' and the label was often used in planning discussions by the Staff. He was awarded a Military Cross for this action.

(As a student of Victoria Cross awards I have always

wondered why Jacka was not given another VC; his Pozières exploit exceeded in bravery, leadership and military result many deeds for which the VC was awarded.)

On that crowded Pozières summit the three Australian divisions engaged lost 23,000 officers and men in less than seven weeks. The windmill site, bought later by the Australian War Memorial Board, with the old mound still there, marks a ridge 'more densely sewn with Australian sacrifice than any other place on earth'. In those forty-five days Australians launched nineteen attacks, all except two being on a narrow front, sixteen at night. They knew their constant advance, during a time of deadlock, would compare with any other achievement on the Somme. Under bombardment of intensity and duration probably unsurpassed, they had held every trench once firmly captured. But they felt little confidence in the high tactics and strategy of it all.

The *Official History* notes that Pozières had many results besides the direct one of shaking the Germans in the key position in the first battle on the Somme. It brought the main Australian forces into the whitest heat of modern war. All the Australian divisions in France were afterwards classed by the British High Command as among those tried British and dominion divisions upon which any responsibility could be placed. Haig surprised Queen Victoria's son, the old Duke of Connaught, by telling him that the Australians were among the best disciplined troops in France. 'When they are ordered to attack they always do so,' he said.

Up till that time the supreme punishment in the AIF had been to be discharged from the force in disgrace. After the dreadful bombardments at Pozières that punishment for some types of men in the force had little effect. Return to Australia was no longer any deterrent for the persistent deserter. Absence without leave increased. At times of strain or even before a great battle, the very time when the average Australian refused to go sick or not infrequently broke away from his convalescence to get back to his mates in the line, a certain section persistently went absent. In almost every other army such desertion could be punished by death, but by the Australian Defence Act this punishment was restricted to cases of mutiny and desertion to the enemy. The restriction was fully supported both in Australia and in the services; the general feeling was steadily against the infliction of the death penalty on men who had volunteered to fight in a cause not primarily their own. Consequently the AIF had to rely increasingly on the leadership and example of its officers and NCOs. One penalty included the publication of lists of offenders in the Australian newspapers.

Absence without leave did not mean that any Digger was afraid of facing danger. These men could take danger in their stride, death didn't frighten them. But they could see no point in useless wasteful sacrifice; they objected to being 'thrown in', as the British communiqués put it, like cement into a mixer. They wanted to know what was going on and just how their actions would help. When it was patently obvious that they were being forced to fight merely because British High Command thought it was about time there was another attack, without possibility of any real advantage, they didn't want to be 'in it'.

English soldiers might have then accepted the creed of 'theirs not to reason why', but Australian troops must know why.

After Pozières the AIF was asked to vote on the conscription issue—one of Australia's greatest political upheavals of the century. Should men be compelled to join the army and be drafted to the war? It was confidently expected that the Diggers would support conscription to a man; after all, they were always grousing about stay-at-homes and 'yellow-livered bastards' who wouldn't do their bit. They often referred contemptuously to the 'would-to-Godders'; the term came from 'Would to God that I could go to the war'.

The vote did not work that way. The back-line and base area troops—supply men, bakers, mechanics, workshop soldiers and the others—did vote heavily for conscription. But most of the infantry and gunners, the pioneers and the medical corps men, voted against conscription. Many of them said, 'I wouldn't bring my worst enemy into this bloody mess!'

This is what they said but it was not the real reason for

voting against conscription. After Fromelles and Pozières and their aggressive patrolling on the Western Front generally, these men, all volunteers, were an élite. They had built up a brotherhood with only one qualification for membership— willingness to have a go. The brotherhood was bound together by the acceptance of self-sacrifice and suffering. The Diggers did not want that brotherhood defiled by dragging in unwilling conscripts. They were asked about conscription again in 1917. And again, practically half the men voted against it. In the opinion of most officers the vast majority of the frontline soldiers hardly needed to think about the issue.

The 1st and 4th Divisions of the AIF were sent to the Ypres Salient for a rest in the latter part of 1916, though the rest was punctuated by numerous small raids. The front was fairly quiet at that time and the Diggers, though they generally liked the raiding, were glad that no major battle seemed to be contemplated.

But they soon found themselves back on the Somme and its great seas of mud and misery. The least effort was exhausting and fighting was practically impossible. At times the Australians and Germans stood on their respective parapets without firing; their rifles were choked with mud. Horses and pack mules which became stuck in the mud had to be shot; men were pulled out but often the mud kept their boots and trousers.

My mother, with the 3rd AGH at Abbeville, nursed men who were exhausted to the point of heart strain and stupefaction. As on Lemnos she encountered hundreds of cases of trench feet; this ailment resulted from local stoppage of circulation and often ended in gangrene and the amputation of feet. She read a report from the Adjutant-General's department of the Fourth Army which stated decidedly that the 'trench feet problem is largely a matter of discipline'. The Staff officers, who were never in the mud, could not understand why the men were not made to wear loose sandbags around their legs and occasionally rub their feet with whale oil and put on dry socks and boots 'in specially provided drying places'. They should have an occasional drink of hot coffee or cocoa. My mother said that this edict made the hospital staff 'furious'. There was neither whale oil, nor dry boots and socks nor 'drying places' and the soldiers were too exhausted to do more than squat or stand miserably in the mud.

In November 1916 Diggers took part in the battle of Flers under dreadful conditions. Charles Bean records that one or two young soldiers, broken by their privations, walked over to the enemy. Another, told that his battalion was ordered back into the line, said, 'I'm not going in—I'm finished,' and shot himself in front of his mates.

The Somme battle ended with 500,000 British casualties— the figure which Haig had budgeted for. But some action continued as Haig wanted to keep the enemy 'under strain' during the winter. Just which side was under the greater strain is arguable but the Diggers did as they were asked and engaged in many 'minor operations'.

The Germans made a controlled withdrawal and on 17 March 1917 soldiers of the 5th Division occupied Bapaume without much of a fight. The Diggers' spirits, always mercurial, began to rise and one boy wrote home to his family, 'I think we have Jerry on the run. Now I'll be glad when spring comes.'

But Jerry wasn't running, winter was still sitting over the battlefields and the Digger had not yet finished building his reputation.

9

'With Proud Deliberation'

South-east of Arras, at Bullecourt, the Germans had built an immensely strong fortress with concrete machine-gun emplacements, cellars and a deep tunnel through which reserves were rushed to the post. Bullecourt consisted of about thirty cottages and a refinery and a large brick building. A strong trench system linked village, fort and the low hill which gave the Germans some dominance.

On 10–11 April the 4th and 12th Brigades of the 4th Division were committed to a rash and ill-planned attack with tank support. Despite Australian protests about the inadequacy of the preparations to attack a major enemy position, they were told that the assault was to be made and pressed hard. At the time Albert Jacka was Intelligence Officer of the 14th Battalion and was given the dangerous job of making a night reconnaissance of the thick barbed wire in two places and went out again to lay the white tapes for the jumping off line. When two Germans approached Jacka pressed the trigger of his revolver but it failed to fire. The Germans could not be allowed to get away as they had seen so much so Jacka rushed them, seizing the officer first and then the soldier. He brought them in as prisoners. Only minutes later the 4th and 12th Brigades assembled on the tapes, ready for the attack. But for Jacka's quick, brave action they would have been shelled and machine-gunned. He was given a second MC.

In the attack at Bullecourt most of the tanks and some of the crews failed and the artillery left enemy wire uncut. The assault was too far committed when the tanks broke down and it was left to junior officers' leadership. Because of misleading reports the artillery had stopped their fire too soon and then they fired on their own infantry.

One of the most outstanding officers was Major Percy Black, a former Western Australian gold prospector. 'I may not come back,' he told his CO, 'but I will get to the Hindenburg Line.'

Major Black led through the storm of bullets to the wire of the first trench. The Germans, Württembergers of the 27th Division and good troops, had been shaken by the appearance of the British tanks and fled to a support trench 180 metres farther back with a belt of intact wire in between. Major Black found an opening and was putting his men through it when he was killed. With such inspired leading the Australians took all their first objectives. But now, because of lack of artillery support, they were left isolated and some hours later the Germans on all sides were able to move and shoot with impunity. Eventually a senior officer called a conference and it was decided, if necessary, to fall back into shell-holes near by. By 11.30 a.m. it was clear that the Diggers must try to return to the Australian lines. Terrible fire was sweeping the ground but Captain Harry Murray told his men, 'It is either capture or go into that.' Very many tried. Murray was among the comparatively few who got through. Eventually the 48th Battalion and part of the 47th were completely cut off. At that time the 48th was commanded by Lieutenant-Colonel Ray Leane. Under his leadership the 48th turned and in fierce fighting took the trench behind it.

The six and a half battalions and support units engaged lost more than 3000 officers and men. The 4th Brigade suffered most severely, losing 2339. Survivors of the 48th Battalion—which was to become illustrious—came out an hour after every other battalion, under murderous machine-gun and rifle fire, 'but with proud deliberation and studied nonchalance, at walking pace ... carefully helping the wounded and with their officers bringing up the rear. Wherever Australians fought, that characteristic gait was noted by friend and foe alike but never did it furnish such a spectacle as here', noted the official historian.

Australians have never liked being pushed out of positions and whenever this has happened they become more dangerous than ever, for their pride is stung and they are apt to retaliate violently. This is what happened a few days later, at Lagnicourt on 15 April, when Prussians pushed Australians out of the village and reached some field guns. Without waiting for orders the Australians, mostly Diggers from Queensland and New South Wales, counter-attacked with the bayonet. The Prussians, all members of famous Guards divisions, fell in wild disorder and by 8 a.m. were in full retreat.

Australian bayonets were so close and the Diggers so annoyed that the Prussians became confused and could not find the openings to their own wire. They were shot down in hundreds and another 330 surrendered.

'The bastards!' a Digger growled as he returned to his trench in Lagnicourt. 'It's Sunday and I was having a nap.'

Diggers were reluctant to accept capture and records contain many instances of dramatic escapes. In a way, the escape of Corporal C. W. Lane and Private R. C. Ruschpler of the 24th Machine-Gun Company was not so very dramatic, but it was enterprising and unusual.

Captured at Dernancourt, Lane helped a wounded Australian, and Ruschpler a wounded German, to a German advanced aid post. They made themselves useful in the aid post with the idea of escape after dark. The German doctor in charge told them to take a German officer and an Australian to a cellar, where they would be safer from the barrage then falling. Leaving the room, the two Australians narrowly missed death from the British shell which burst in the post, decapitating the doctor.

About four in the afternoon they were detailed to carry the doctor's body to his billet in the village of Meaulta. Escorted by two unarmed Germans, the Diggers arrived at the billet to find the doctor's dinner ready for him on the table, so they ate it. The German escort helped to polish it off, but apparently they were reluctant to do so. Perhaps they weren't so hungry as the Diggers.

The two Australians spent several days in Péronne and were then sent to Bray to work on an aerodrome, within reach of British shellfire. A shell blew a hole in the wire around the drome and one night Lane and Ruschpler got through it and made for the British lines. They sneaked between two German sentries, crossed no-man's-land during a barrage and finally reached the British lines, where they had their narrowest escape of all—they were nearly shot by sentries.

On 3 May a second attempt was made to capture Bullecourt and units from the 1st, 2nd and 5th Divisions were engaged. On the 3 and 4 May the Australians were so hard-pressed that they were close to returning to the starting-line, as had happened on other parts of the battle line. Their hold on the Hindenburg Line was slender. Originally it was a mere 365-metre break made by the 23rd and 24th Battalions (Victorian), whose third wave had passed on towards Riencourt before the failure of the attacks on Bullecourt had been realised. The 5th Brigade (New South Wales) which had advanced on the night of the 6th, forming the extreme right flank in the order of battle, had found the German lines held in great strength, and all those who reached it were killed.

Part of the brigade later joined their Victorian comrades, and with great dash bombed down the German trenches towards their first and second objectives. This work was continued by the 7th and 1st Brigades, and by night-time the Australians had secured all that portion of the Hindenburg Line marked out for them in the general scheme—1100 metres. More than half of this had been won by bombing, which was now the intensest form of hand-to-hand fighting on the Western Front. Some Western Australian troops had also been sent against the south-western side of Bullecourt, to aid the 62nd Division; the first wave was annihilated, and the orders to the others were countermanded.

Throughout the night the Germans tried desperately to turn the Australians out of the line, and counter-attacks were numerous. The Australian position was like a large flower on a very slender stalk—a single communication sap, dug by the engineers during the first hours of the attack, being the only link between the new positions and the old. The heaviest counter-attack was made at 10 p.m. and consisted of waves of 'storm troops' who advanced from Bullecourt on the one side and from Queant on the other. They used flame-throwers, mortars and bombs, and were met with a hail of Stokes' mortar-bombs and with cold steel.

The Australians' right was slowly driven in. The Germans reached even to the sap. They came on wave after wave; the heroic survivors of the 23rd and 24th Battalions, which still clung to their gains of the morning, seemed doomed to isolation. 'The precious grip on the Hindenburg Line,' wrote an Australian correspondent, 'seemed to slacken and fail under mere weight of the enemy thrusts.' Back at the railway embankment, the old Australian front-line, every man was given a post of defence. The brigadier seized a rifle. About 700 metres forward in the new line the word went round to retire.

' "Who said retire?" the men asked, "None of *our* officers will say retire." They resolved, these Victorians, to die where they stood rather than give up their gains. And it seemed at that moment that the choice had definitely come' (*The Times*).

The counter-attacks were beaten back before midnight, and during the day, troops of the 1st Division recovered by bombing all the lost ground. By the evening of 4 May the battle had become a struggle for the retention of this pathway through the Hindenburg Line. To the north the fighting simmered down; the hope of great captures was abandoned. But here was the vital breach, through which further advance might become possible.

General Gough brought up the British 7th Division, which relieved the British 62nd on the Bullecourt front; the remaining brigades of the 1st Australian Division moved up in support of the 2nd Australian Division and General Hobbs's 5th Australian Division was brought within striking distance. It was determined to take Bullecourt by a series of frontal assaults, and to hold at all costs the breach in the Hindenburg Line to the right, despite the mass of artillery which the Germans were now concentrating on this solitary spot.

A strong assault was launched by the Gordons of the 7th Division in the early morning of Monday, 7 May. The 207th German Division had been brought up to defend it, and the fighting was stubborn. The Gordons penetrated into the ruins, and at the same time troops of the 1st Australian Division began to bomb down the trenches on the western side. Since 3 May the Australian position had been fully exposed on each flank—the points where their occupation of the German system ended being marked only by sandbag barricades. The Scottish troops clung to a line across the south-eastern corner of the village, and about noon that day the union of Scots and Australians took place in the Hindenburg Line on the south-western slope of Bullecourt, and a continuous front was established firmly from the pounded hillside to which the Australians had so tenaciously held.

The Germans would not yet admit defeat. An Australian general described a counter-attack in which shell-holes were used:

> It was for all the world like a school of seals. First the heads of a number of Germans were seen in the sunken road, near Riencourt, to which some of our men had penetrated during the first minutes of the assault. The counter-attacking troops were forming up. Then they came over the top. They came, two or three hundred together, diving from shell-hole to shell-hole—crawling from one, and seeming to lurch forward and plunge into the next. It was well done, but it was irresistibly funny to watch. Our men stood on the parapet, and breast-high against it, with cigarettes in their mouths, and shot as they have seldom had the chance to shoot. The attackers were simply wiped out with rifle and machine-gun fire, though some got close to our line. They tried at the same time a bombing attack on the flank, and this was well countered by our mortars.

By all the theory of war the Australians should have been thrown out of their position. A captured Prussian officer, who could not understand their venturing to retain so exposed a salient, spoke of them hopelessly as 'those madmen from the Antipodes'. But every metre gained in Bullecourt increased the area over which the Germans had to distribute their shells, and the linking up with the 7th Division firmly secured the left flank.

By 13 May after several days of fighting, the gap in the

Hindenburg Line became three kilometres wide. A final effort was made by Prince Rupprecht to re-establish it. On 12 May he had withdrawn the Lehr Regiment (the 'Cockchafers') from the 3rd Guard Division, which opposed the Australians. (The Lehr Regiment consisted of small detachments brought from the various Prussian regiments to be trained together so as to ensure when they returned to their units that they impart instruction on identical lines.)

The regiment, one of the most famous in the German Army, was told that the honour of recovering the Hindenburg Line was to belong to it, and that after the battle it would be sent to a pleasant resting-place. They rehearsed the attack with great thoroughness. Aerial photographs were taken of the Australian positions, and model trenches made for the rehearsals. The regiment went over the attack by day, and then by night. Little white screens were used to mark the distances, so that the men would by practice know almost by instinct the places they had reached. Every man was taught his exact duty in the attack.

A great bombardment preceded this assault. All day on 14 May German artillery and mortars pounded the Australian line. At night the bombardment intensified, and an hour before dawn it became terrific. At 3.45 the Lehr Regiment advanced. They attacked the Australians from right flank while other troops advanced towards the British in Bullecourt itself. Australians and Germans were only thirty-five metres apart and soon severe hand-to-hand fighting took place.

Bean wrote:

One after another, four waves of dark figures attempted to rush over the tumbled earthen sea against the two ends of the trenches held by the Australians. A good part of them were mown down at once with bombs and machine-guns. A portion managed to struggle through towards our front trench, and the dark figures could be seen running along it and dropping in. But the attack was always utterly disorganised. Within two minutes of the assault the results of all this careful planning and practice had been thrown to the winds. All that remained of it was between two and three hundred Germans in a section of Australian

trench, with scarcely any idea of where they were and
what was happening, machine-gun bullets sweeping above
their heads and making any sort of movement utterly
perilous.

The Germans held their small gain for three hours. None
escaped. All were immediately cut off from their own line
by a heavy barrage, which thundered down behind them.
Two counter-attacks, both launched straight at them across
the top by the New South Wales troops, accounted for the
lot. The first counter-attack drove them into a small corner
of the trenches; the second cleared them all up. It marked
the finish of the German resistance in the battle of Bullecourt.

The prolonged battle for Bullecourt had its main value in
the distinct beating it inflicted upon Prince Rupprecht's army.
Possession of Bullecourt was not made use of in further
movement in this sector. Changes were made in the Allied
plan and the centre of the British actions moved farther to
the north. But Bullecourt tied German divisions to the sector,
it mauled them, and it had a distinct moral effect.

Also, it showed that the very best of German troops could
not win a hand-to-hand fight with Australians.

Every day from 5 May till 17 May the Bullecourt fighting
in which the Australians took the major part was mentioned
in the British bulletins, often the chief item. A French journalist announced: 'The Australians have again captured the
British communiqué.'

The winning of their impossible position in the Hindenburg Line and the holding of it despite seven general counter-
attacks and a dozen minor ones, was, as Haig wrote, among
the great deeds of the war.

Second Bullecourt cost the AIF 7000 casualties. The only
good thing that came out of it was a long and happy rest for
the Diggers. And the plaudits of history. *The Times History*
states, '. . . In the lonely country homes and thriving cities
of Australia, Bullecourt is known with Gallipoli, Pozières
and Passchendaele as a national battle name.'

But, immediately after Bullecourt, most Diggers had never
heard of Passchendaele.

10

'Another Poor Bastard'

By June 1917 the Diggers were in Flanders from which base the Ypres salient bulged into the German lines. It was a flat area and virtually without features, a mass of churned up mud. The slight heights, such as Hill 60—that is, 60 feet above sea level—assumed the prominence that real mountains would have in other areas.

That summer the British plan was to capture the Messines Ridge. It was only about twenty-four metres high but it overlooked the British trenches south of Ypres. Miners had burrowed tunnels from the east and put in place twenty-one great mines which were to blow the entrenched Germans off the ridge. The Australian 3rd Division, which had not before 'hopped over', was on the southern flank of the attack line. It would go in after the explosion and the 4th Division would follow it in the afternoon.

As always some of the men had premonitions of death; one who appeared to have such a premonition was Pte G. H. J. Davies, a clergyman fighting in the ranks. 'I am ready for "the big push",' he wrote. '. . . If I die I give my life willingly for my country, knowing that it is given in a righteous cause . . .' He confessed that he hated war and 'the curse of military life' but as was taking part in it he would 'fight like only one facing death can fight'.

This seems to have been the attitude of many Diggers at that time. They had what my father called 'a fierce resignation'. He meant that the men, resigned to the fact that they could not avoid battle, were determined to fight savagely.

On 7 June, nineteen of the mines (another two did not explode) blew Messines Ridge into a different shape. The shocked German survivors were shelled but they bravely stood up to the infantry assault. The Australians gained three kilometres of territory—nevertheless 6800 were killed or wounded. New Zealanders and British troops performed equally well and Messines was instantly recognised as the greatest British victory of the war to that time.

Then came Haig's great but ill-considered attempt to break through the entire German defence system from the Ypres salient with a massive artillery and infantry assault towards Broodseinde and Passchendaele Ridge. The Australians had their first victory on 20 September against Menin Road Ridge, then another on 26 September at Polygon Wood. The fighting at Broodseinde on 4 October resulted in a further victory—achieved with great gallantry. It was even more decisive but the cost to individuals and the AIF was fearful. 'I feel about 80 years old,' a Digger wrote afterwards. But at least he was alive; in the Flanders battles so far the AIF had lost nearly 17,000 men.

At this point the battlefield became a quagmire from rain and with the under-surface water table destroyed by incessant shellfire. The breakthrough became impossible and the most that Haig could now hope for was the capture of the ruins of Passchendaele, which was achieved largely by four Anzac divisions attacking side by side—the only time this ever happened—on 6 November. In eight weeks they had suffered 38,000 casualties.

After what became known as the Battle of Passchendaele the Diggers were quiet, disillusioned and hoping for a wound which would get them evacuated to England—'a blighty one' they called it. Some men actively sought a wound by exposing an arm; some walked about on top of a parapet. The 'fierce joy of battle' had evaporated. 'Doing a hop-over' was no longer stimulating but something to be endured with gritted teeth, except among the reinforcements who were new to the front, fresh and eager to prove themselves to the veterans. Survival was uppermost in most minds, together with the conscious decision to die 'with good grace', as one soldier expressed it. The nurses knew at this time that some of the wounds they were dealing with were self-inflicted but they said nothing to the soldiers concerned; they read the haunted horror in their eyes.

Having won some notable victories the Diggers were re-warded—by being kept in the trenches, or what passed for trenches. Actually, they were channels of oozing mud into which many men had disappeared. Any kind of movement was a tremendous effort and working parties took hours to cover even a short distance with their heavy loads. Horses, carts and even great guns vanished into the mud-sea. 'It isn't right for Australians to be in a place like this,' one soldier said.

Rule vividly describes the winter of 1917:

Winter set in, the like of which had not been experienced for twenty years. The cold was so intense that sleep came only in snatches. Our feet were the worst problem. It was utterly impossible to get them warm because our one pair of boots was continually getting wet, and the work of getting into them after getting out of our three blankets was a painful ordeal. The boots were frozen stiff and were like boards. Many of the lads on rising looked liked old men crippled with rheumatism and it was painful to get about.

At daylight every morning I had to get out and assemble my bombers for fatigue duty. It was difficult to induce the poor wretches to leave their blankets for the frozen boots. When the command 'Fall In!' was given it was not unusual to see men lining up with a piece of bread and bacon in their hands. It would be impossible to convey to civilians the misery and squalor of the whole thing. In the bleak wintry days, out on the job of road-making or whatever it might be, men would turn their backs on the sleet and rain, just as animals do, and wait for the storm to pass. The day's work over, back to camp for their tea and stew, and, when darkness began to fall, to huddle up in their damp overcoats under a sheet of iron and get what sleep they could. Men made the best of it and tried to be cheerful, and occasionally one would hear bursts of laugh-ter, especially if the sun favoured us with a visit through the clouds.

The laughter was sometimes the result of some new bit

of black humour passing through the camps and trenches. One story went like this:

A Jerry spy, captured in the lines and sentenced to be shot, was being escorted through five kilometres of Flanders mud to have the sentence carried out. Very fed up, Jerry complained loudly: 'Vot for take me through mud to be shod? Vy nodt shoodt me and be with it done?'

The Digger, also very fed up, said, 'You've got a lot to complain of, you have! I've got to walk back through it!'

Another story which began at this period—and it could be true—concerns Scotty, an excellent soldier but the bane of authority. He was always in trouble with the NCOs and his paybook, as a result, had a lot of entries in red—indicating fines imposed by his OC or CO.

The Company Quarter-Master Sergeant died and somebody suggested that a suitable memorial should be put over his grave. The company Diggers were asked for contributions.

A sergeant approached Scotty and said, 'Will you give five francs to bury the Quarter bloke, Scotty?'

'Certainly, take ten francs,' Scotty said.

The sergeant was startled by this unexpected and uncharacteristic generosity. 'But the others are only giving five,' he said.

'Aw, keep the ten,' Scotty replied. 'And bury the bloody sergeant-major with him.'

By this time, towards the end of 1917, the term 'Digger' was in general use, both for the Australians and New Zealanders. The men themselves called one another 'Digger' or 'Dig'—the inevitable Australian contraction. In truth, they did a lot of digging—mostly graves for departed friends. The last service a man could perform for his mate was to fashion a cross and stick it in the ground above the grave. One of my father's friends told me during the 1930s that he invariably put the letters A. P. B. after each dead man's name, rather as if they indicated a decoration. They stood for 'Another Poor Bastard'.

At this time, in mid-winter, the first lot of 'Six-bob-a-day tourists' travelled through Italy and were dubbed the 'Italian

Anzacs'. When these men were landed at Taranto, in the south, many rumours quickly spread. It was generally agreed that they were there to help the Italians who were suffering greatly under Austrian pressure. But they were put on a train for France and after a journey of eight days untrucked at Cherbourg, having left behind them men at every stopping place en route—all suffering from frostbite. When they reached the battle areas these Diggers were deeply aggrieved when told, 'Wait until you feel the cold here!'

Reinforcements, who often outnumbered the veterans, listened intently to the stories told by the men with longer service, and sometimes they unwittingly provided a laugh. In one battalion, Darkey, the unit wit, was telling of an experience when he and a pal were squatting and talking in a captured German dugout after an advance. Jerry commenced reprisals and a German shell, which proved to be a dud, came right into the dugout and hit Darkey's pal in the chest.

'Did it kill him?' asked the nearest reinforcement.

'No,' said Darkey, 'only made him cough.'

With native resilience, aided by souvenir hunts and occasional encounters with willing French and Flemish girls in the rest areas, the Diggers pulled themselves out of the despondence which afflicted all the old hands at the end of 1917. They criticised practically everybody, verbally and in their letters: the Staff—how they hated the Staff!—reinforcements, the base area troops, the military police, English officers, British soldiers—though never the Scots—French soldiers, the rations ... It was a kind of therapy. Another kind was a selective forgetfulness, putting out of the mind the foul and bitter memories. With the arrival of 1918 the Diggers were ready again for whatever war could throw at them.

11

'Beaucoup Australiens ici'

During 1918 the Diggers several times proved that besides being good fighters they were also good soldiers in the intellectual sense. Australian commanders at last had a few chances to plan attacks and consequently the Diggers had greater success and fewer casualties. E. J. Rule noted:

Coming back towards Gueudecourt and going down the slope I saw no end of Aussie dead, but they were scattered very much about the slope and it was plain that our tactics were far more effective than the English, who always seemed as if they herded together and so ran great danger of being cut to ribbons.

On the evening of 10 February, Diggers made a successful raid near Warneton, south-east of Messines, and other points received some attention. The raids were made because of the increase in traffic along the Houthem–Comines–Wervicq railway. A prisoner had revealed that his division had been intensively trained for an offensive action and that it had now been brought back for the attack. It was desirable to find out what was being done.

The Australian attack was made on a strongly defended position immediately north of the river Lys. The plan included feints on the left, which kept the enemy in doubt as to the exact front of the raid. The German wire entanglements were cut by trench mortars and artillery. At ten o'clock a heavy barrage was laid down along a considerable length of front, combined with heavy counter-battery fire against enemy guns, and a bombardment of enemy headquarters and dugouts behind Warneton. The attacking force entered the enemy lines over a front of 365 metres, and met a large garrison, which they overcame by bomb and bayonet fighting.

Leaving parties to destroy the dugouts and other works, the assault was then pushed forward against the second line, which was attacked with the bayonet. Trenches behind the second line were also entered, and many dugouts destroyed. After dealing destruction in every possible direction for half an hour the raiders withdrew, having killed in infantry fighting alone ninety Germans. Many more were killed and wounded by artillery fire and Lewis-gun fire during the withdrawal. Australian total casualties equalled only one-sixth of the known enemy losses and they inflicted great damage.

Near Ypres, at a position known as Spoil Bank—the spoil being the waste earth from a railway cutting—a young Australian machine-gun officer gave one of the most famous orders of the war. Towards the end of February 1918 the AIF 3rd Machine-Gun Company was moving into the line and officers were sent forward to reconnoitre the position where the guns were to be posted. They reported back to the officer in charge of the section, Lieut. F. P. Bethune, that it had a field of fire of only 5½ metres and was useless; the crews would be shot before they could get into action. Bethune protested to his CO but the order stood; as a matter of honour Bethune insisted on being the officer in charge of the post. He called for volunteers and every man stepped forward; he selected three veterans and three new men.

He was overtaken by a runner who told him that orders had changed; he was to post his guns at Bluff Bank. Bethune got his guns into position, loaded all ammunition belts and placed the 10,000 rounds per gun ready to hand. Nearby infantry was moved back to prepare for an attack and Bethune's position was dangerously exposed. He considered that each man should have written orders 'so that if a man had to die, he should die in his own lighthearted fashion, in goodly company'. This was his order:

SPECIAL ORDERS NO. 1 Section 13.3.18
(1) This position will be held, and the Section will remain here until relieved.

(2) The enemy cannot be allowed to interfere with this programme.

(3) If the Section cannot remain here alive, it will remain here dead, but in any case it will remain here.

(4) Should any man through shell shock or other cause attempt to surrender, he will remain here dead.

(5) Should all guns be blown out the Section will use Mills Grenades and other novelties.

(6) Finally, the position as stated, will be held.

E. P. Bethune, Lieut.
O/C No. 1 Section

The position was held and the crew survived, though some were killed later. Bethune's battle order was circulated by HQ of the Australian 1st Division and by other staffs. In the American forces copies of the order were mimeographed and distributed as an admirable model of all that standing trench orders should be.

Bethune won the MC. He said in 1937: 'My men knew that I knew they could not consider such a possibility of surrender and so between us we enjoyed in silence the joke that to an outsider might have seemed a little grim.' A clergyman in Australia before the war, Bethune returned to the cloth. The story of his order is incorrectly told in several books, including *Australia in Nine Wars*. Here the order is said to have been issued 'by a junior officer on the Hazebrouck front and found on him and the bodies of his men'. As can be seen, Lieut. Bethune survived. Perhaps copies of the order were carried into action at Hazebrouck by other AIF men.

During the winter of 1917–18 the AIF had 117,000 troops in France and for quite a long time their main activity was patrolling. The Germans, usually employing more formal methods than the Australians, made fifty-four recorded attempts in which they obtained prisoners, or killed and identified their opponents on ten occasions. They lost prisoners or were themselves identified on forty-two occasions. The Australians in twenty-five recorded attempts secured prisoners or identification on fourteen occasions, and left them on only seven. One German regimental history states that the patrol

enterprises of its regiment, though keen, brought irreplace-
able losses. The information obtained was often not worth
the cost. Another history says: 'The enemy infantry was
always on its guard.' Yet, ever since Gallipoli days, the
Australian infantry left the illumination of no-man's-land by
flares almost entirely to their opponents. They preferred the
dark.

In early March 1918 a great attack was expected daily and
the nervous strain on the troops was severe, but it was notable
that there was no sign of lack of confidence among the
Diggers. On 21 March the Australians, ninety-seven kilometres north in Flanders, heard of the German onslaught
which had broken through the Fifth Army Corps and compelled it to retreat. On 25 March it was stated in Amiens
that there was possibly no British division between that town
and the Germans, forty-five kilometres away. Also came word
that some big German guns were shelling Paris from a forest
113 kilometres distant. In this crisis the Australian infantry
divisions began to strain on the leash which kept them idle
in the north. An infantry officer wrote: 'We all have the
feeling that if we could only get down there it would be all
right.'

On 25 March the 3rd and 4th Divisions were on their
way, their bands playing 'Colonel Bogey' or the men singing
the old marching songs. This time they could see that every
step led directly towards beating the Germans.

At Saint-Pol, fifty-five kilometres north of Amiens, where
no Australian infantry had ever been stationed before, they
were immediately recognised and people began to call to one
another, 'Les Australiens'. The old people already on their
way out of the danger zone began to unload their carts.

'It is now not necessary,' one of them told an Australian
who asked the reason.

Passing through Corbie on the way to the battle-area an
Australian officer heard what he said (in 1933) was the best
compliment ever bestowed on the Diggers.

An English brass hat was ordering the civilians to leave
quickly and join the remnants of the retreating Fifth Army.
Two old ladies were painfully wheeling their goods away in

a hand-cart when 'with a roar, a curse and a ribald song' the 16th Battalion (Western Australia) swung round the corner.

One old lady squealed. '*Soldat Australie?*'

'*Oui, Madame,*' the young officer assured her.

She said, '*Tres bon, fini la Boche.*' With a rude gesture to the brass hat she turned the cart round and headed for home again.

The 4th Brigade, apart from the rest of the 4th Division, was ordered to stay at Hebuterne, and there repelled several German attacks to break through. Not until a month later did the commander of the British IV Corps, as his messages made clear, entrust the position to the other available troops.

On 27 March the 3rd Division, taking up positions on the Amiens–Albert road, found that they were the only traffic heading towards the enemy. Streaming past them in the opposite direction were villagers and British soldiers, including heavy artillery. They shouted to the Australians that they were going the wrong way, the Jerry would souvenir them.

An artillery brigadier said to Lieutenant-Colonel J. D. Lavarack: 'You Australians think you can do anything, but you haven't a chance of holding them.'

Lavarack said: 'Will you stay and support us if we do?' and this the brigadier willingly did. Again and again the Diggers were told by passersby: 'You can't hold them.' Some Australian leaders were a little anxious as to how all this advice and the depressing sights would affect the men. They need have had no worry. The troops were overflowing with confidence. A British major of artillery said: 'They were the first cheerful, stubborn people we had met in the retreat.'

In the French villages wherever, during those weeks, the Australian battalions marched in, they were met by striking demonstrations of affection and trust, and this too reacted strongly on them.

'*Fini retreat, madame,*' said a Digger to a French woman as he sat cleaning his rifle while the 3rd Division halted in Heilly on its way to the triangle between the Ancre and the Somme. '*Fini retreat. Beaucoup Australiens ici.*'

Men and officers were shocked by some of the scenes on the roads. . . . The panic near Hebuterne, a car with rattled

TRALIANS PARADING FOR THE TRENCHES. 80.

OFFICIAL PHOTOGRAPH, CROWN COPYRIGHT RESERVED.

The original caption for this *Daily Mail* postcard notes: 'These are the men who shortly after midnight on Sunday 23 July 1916 took Pozières by a splendidly dashing advance through shrapnel, shell and machine-gun fire.' Probably half of these men were killed in that advance.

Diggers of the 24th Battalion resting in the safest place they knew, a mine crater behind Albert Redoubt. The photograph was taken on 21 September 1917. (*Imperial War Museum*)

Troops of the 45th Battalion in a supply trench called Garter Point near Ypres, Flanders, on 29 September 1917. Some of the men are wearing souvenired German caps. (*Imperial War Museum*)

A characteristic group of Diggers on a General Service wagon passing through a village near Amiens, France in 1917. (*Imperial War Museum*)

staff officers too hastily retiring, parties of men without arms, stragglers who had lost their units . . . destruction in villages on the Ancre.

At 11 a.m. on 27 March two battalions of the 3rd Division relieved the remaining completely exhausted British infantry in the triangle between the rivers. Two battalions of the 4th Division were ordered to move up in close support of the exhausted English 9th Division. The Australians advancing along the open hill-top astride the straight avenue of the Amiens–Albert road were shelled heavily. At dusk came the order to relieve the 9th Division (Scots) at the railway round the foot of the hill.

After seven days' constant fighting the Scots were exhausted. 'Thank God!' they said when they saw the Australians. 'You'll hold him.'

Three kilometres to the south-west the 3rd Division also was in the front-line. The overall situation became serious when Marshal Pétain decided to abandon the British flank and withdraw to cover Paris. The position was desperate until Haig asked the French Government to appoint Marshal Foch 'or some general who would fight' to take over supreme command of the operations in France.

But by a mischance the Third Army was withdrawn, so that on the night of 27 March the two Australian divisions, with the remains of the English 35th Division between them, became responsible for the Third Army's flank. Early next morning, Sergeant S. R. McDougall, 47th Battalion, watching a level crossing on the railway, heard marching troops approaching through the dense mist.

McDougall ran along the railway to alarm the nearest outpost. The leading Germans threw bombs over the embankment and hit a Lewis-gun crew. McDougall grabbed the gun, shot the leading Germans crossing the railway and then sprayed the others along their side of the embankment. The attack spread along the 12th Brigade's front, but was beaten back with heavy loss to the Germans. The only ones to get across the line were captured by McDougall, who won the VC for his exploit.

On 30 March a fresh German division tried to push back

the Australian 3rd, but suffered the 'worst miscarriage in its existence', as a German regimental diary described it. Finally the German attack was halted by a battalion of the 9th Brigade (being held in reserve) under Lieutenant-Colonel Morshead (to become twenty-five years later the famous leader of the Rats of Tobruk).

Without artillery, but with the help of the 12th Lancers, the battalion was to re-establish the badly dented British line. Except for the 29th Division on Gallipoli, the British cavalry, whom the Australians watched with admiration throughout this campaign, were almost the only troops of the old pre-war British Army with whom the war brought them into contact, and the co-operation was always enthusiastic. On this occasion, though most of the desired ground was not gained, the aggressiveness of the cavalry and the Australians helped to halt the German offensive.

On 4 April the Germans hit again—this time with fifteen divisions on a front of thirty-four kilometres. A threatened puncture of the line was again stopped by the cavalry, this time helped by the 33rd Battalion. Bridges on the Somme were guarded by the 15th Brigade and some of the brigade's officers and men crossed the river to help rally the British infantry and hold the line. In mid-afternoon the Australians again saved the day.

Attacking strongly, the Germans drove back part of the 18th British Division, south of the Australians. The 9th Brigade's southern flank had to be swung back to avoid envelopment and soon the whole line retired; the Germans reached the outskirts of Villers-Bretonneux and the township was as good as lost.

At this moment, the 36th Battalion (under Lieutenant-Colonel Milne), waiting in a hollow near the town, made a spectacular charge, and British remnants joined in. The Germans saw the even line of advancing Australians, their bayonets flashing, hesitated unevenly, stood apparently fascinated for a second or two, then bolted back 1½ kilometres to the old trenches. Australian bayonets had won the day. Neither then nor at any time since has an enemy been over-eager to tangle in a bayonet fight with an Australian. An Australian bayonet charge always advanced in an even steady line, a much more frightening sight than a ragged charge with men spread over a depth of fifty metres or so.

The strongest attack made against Australian troops occurred on 5 April when the 12th Brigade and soon after the 13th Brigade (both 4th Division) were attacked near Dernancourt. After a fierce fight in the morning mist, the Germans forced their way through the Australian line and attacked some posts in the rear. They pushed up a hill, outflanked the 12th Brigade's supports half-way to the top and established themselves, with a field gun, behind the outpost line of the 48th Battalion.

The 48th had been ordered to hold the front-line at all costs, but at noon, with complete envelopment certain, the senior officer present, Captain F. Anderson, ordered withdrawal. The 48th's tradition, established a year before at Bullecourt, now showed up brilliantly. Calmly, unhurriedly, the battalion came out of its impossible situation. Some of the posts were overrun, but the men in them fought to the end. German soldiers erected two wooden crosses at the site of one outpost, each of them marked in pencil, 'Here lies a brave fighter.'

The day was not lost. At 5.15 p.m. reserves of the 12th and 13th Brigades advanced over the hill and although they met intense fire they drove the Germans part of the way down the hillside and held them there. The two Australian brigades lost 1000 men between them and the three attacking German divisions about 1800.

Senior Australian commanders, like their juniors, have never been willing to accept an order without question when they believe it to be wrong. A notable instance of this occurred on 24 April 1918, when Brigadier-General Glasgow, commanding the 13th Brigade, was told by his British superior that he must attack German positions at Villers-Bretonneux at daylight on 25 April and from the south northwards.

Glasgow said: 'If God Almighty gave the order we couldn't do it by daylight.' He proposed to attack at 10.30. Also, he said, he wanted to attack eastwards, by surprise, and without

previous bombardment. The matter went to the corps commander several times—he must have been very irritated—
and finally Glasgow agreed to attack at ten o'clock.

When the brigade assembled it was seen and fired on.
Advancing, its flank was stopped and for several minutes the
whole attack was jeopardised. This was when a junior commander chose to deviate from his orders. He was Lieutenant
C. W. K. Sadlier of the 51st Battalion. Sadlier was flat on
his stomach sizing up the situation when his platoon sergeant,
Stokes, crawled up to him and urged him to attack the
German machine-gun nests—an action that was not on the
programme.

Sadlier sent a message to his company commander, then
attacked the German machine-gun nests, destroying all six
of them so quickly that the entire delayed flank swept forward
on time with the rest of the assault.

Sadlier won the VC and Stokes the DCM for this audacious
assault. Brigadier-General Glasgow did not reprimand Sadlier
for exceeding his orders; after what the brigadier himself had
done he was hardly in a position to throw his weight about.

The Times History, dealing with this period, says:

> From 27 March until the end of April the Australians were
> in the hottest of the fighting in the Somme valley, and on
> the heights overlooking Amiens from the west. Their
> counter-attack on Villers-Bretonneux on 24 April [the
> incident described above was part of the overall offensive]
> was, perhaps, the most memorable of their doings in this
> fighting, and it marked the end of the German offensive
> towards Amiens. Marshal Foch described this battle as the
> turning point of the German offensive. The Australians,
> he had said on an earlier occasion, were 'shock troops of
> the first order'.

12

Peaceful Penetration

In May 1918 the Australians and New Zealanders began a private war against the Germans. This little war within a war was one of the most remarkable aspects of the whole conflict and it did much to give the Diggers their unique reputation.

The Anzacs were feeling pretty good; they had beaten the Germans whenever they had met and now, while the front waited for another German offensive, they wanted to get into some more action. War, at this time, became something of an exciting if deadly game, as Anzac patrols continually pestered the Germans.

They sent out patrols to ambush German patrols; they raided the enemy's outposts, cut his telephone lines, killed his sentries, and, as one Digger put it, 'We generally made bastards of ourselves.' The bastardry never let up and it was well known that the morale of the German units opposing the Anzacs was at a very low ebb. The Diggers called their tactics 'peaceful penetration', and for four months they penetrated sharply and ruthlessly and not at all peacefully.

During May the Americans started to arrive and almost instantly the Australians resented them. 'Going to win the war for us, Yank?' a Digger would often say. Apparently the Americans were diplomats, for the usual answer was: 'Maybe we will, if we can fight like you guys.'

Because of the humility of the Americans, resentment against them did not develop and there was soon close co-operation and friendship between the AIF and the AEF.

Brigadier-General C. Wagstaff, who spent much time with both forces, said: 'The two armies are the nearest thing possible to one another. Their discipline is founded on the

personal influence of the officer over his men . . . and provided they get the right class of officer there is no trouble
whatever. . . . '

Private war activities had odd results. After a time some
French units of General Mangin's army began to copy the
Anzacs and by July the private war was puzzling German
commanders. It even forced a change of tactics on Ludendorff
himself.

At this time the Diggers were practically filling the official
communiqués and reports by war correspondents. There was
some feeling that the Australians were getting more than
their share of publicity, but the truth was that they were
doing much more than was reported or disclosed. *They were
responsible for almost everything that happened for a full four
months.* And all of it was unofficial—the orders mostly did
not exceed brigade level and often actions were carried out
on the orders of company and platoon commanders acting
on their own initiative.

The first incident occurred on 5 April, when a German
officer with thirty men was sent to penetrate towards Vaire.
On the river flats south of the Somme was an Australian
post, manned by Corporal D. A. Sayers with three privates,
all Victorians of the 58th Battalion. They were protecting a
machine-gun put across the river the previous day on the
advice of Lieutenant H. G. Hanna who could see the chance
of taking the Germans in the flank.

The four Australians watched the thirty Germans until
they were spotted; the Germans began to set up a machine-
gun. Sayers ordered two of his men to fire from the front
while he and the third man crept down a drain leading to
the flank. In a commanding position, Sayers and his mate
opened fire, hitting seven Germans. As the German officer
ordered a retreat Sayers and his men charged, firing from the
hip. Sayers shot the officer. The surviving Germans bolted.
Sayers was awarded the DCM.

This display of initiative by an NCO and intelligent use
of ground was commmonplace in those four months. It was
the type of warfare that suited the Australians' temperament
perfectly. Not for them the blind and senseless attack in the

face of heavy enemy fire. When ordered into this sort of action they went and their record speaks for their success, but left to themselves they could have gained much more ground and lost far fewer men.

In May and June patrols of the 1st Division took German posts almost daily between Strazeele and Merris, in Flanders. By the end of June, Prince Franz, commanding the 4th Bavarian Division, told his men that the Australians' exploits were a disgrace to the division.

Peaceful penetration reached its climax on 11 July, a beautiful summer's day, when the Australians had a field-day's sport with the 13th Reserve Division, after it had relieved the Bavarians. In one efficient stroke after another they captured about 1000 metres of the enemy's front, taking 120 prisoners, including three officers, and eleven machine-guns. The most extraordinary thing was that neither the Allied High Command nor the German High Command knew a thing about the coup, until long after it was over.

On the Somme, too, peaceful penetration went on with the same deadly and, to the Germans, frightening efficiency. On the hot morning of 18 May Lieutenant A. W. Irvine of the 18th Battalion (NSW), guessing that the opposition in a troublesome outpost was asleep, quietly walked across no-man's-land with a few troops, in broad daylight, and brought back alive almost the whole garrison—twenty-one men, a student officer and their machine-gun. No Australian was wounded and the Germans did not know the post was missing until an officer made his rounds that evening.

This was one of the most striking successes of peaceful penetration, and it was not excelled until 1941 when Australians in Tobruk went out one night and bagged an entire Italian company by the same daring method.

According to German records, the commander of the German unit which suffered from Irvine's audacity lost his command by direct order of Ludendorff, who said it was ridiculous for the Australians to be able to do such things.

But Australians still became casualties and at the end of May Captain Albert Jacka was badly gassed. The gas got into some of his old wounds and he was sent to England. It was the end of his military career and 'Jacka's Mob' mourned his departure from the front as if he had been killed in action.

Late in June 1918 Diggers of the 1st Division took part in what the *The Times* called 'a curious little affair'—capturing some hostile posts west of Merris on the Marne, taking forty-three prisoners, nine machine-guns and two mortars.

A patrol of South Australians in no-man's-land about six o'clock noticed that the enemy in front of them had a distinct disinclination to fight. They therefore rushed the post and captured the whole garrison. Some of the prisoners then pointed out the position of the next post, so another patrol took that also. In this impromptu and highly irregular fashion five other enemy posts were taken one after the other. The position was consolidated by the capture of a strong dugout a little to the south of those already taken. This advanced the Allied line about 275 metres, on a front of 800 metres, with hardly any loss to the Diggers. Next day the Germans tried to retake the ground, but they were repulsed with loss.

Peaceful penetration was involved in some big operations, too. Haig asked the Australian corps to seize a large part of a plateau east of Villers-Bretonneux. Elaborate plans were drawn up for the attack, but they were never used, for the good and sufficient reason that the positions were captured by peaceful penetration the day *before* that set down for the attack.

The report of the German 41st Division told the story briefly:

> At 11 a.m. on 8 July the enemy penetrated the forward zone of the 108th Division by means of large patrols without artillery preparation. He occupied the trench where our most advanced outpost lay and apparently captured the garrison, comprising fifteen men. The larger part of the forward zone has been lost.

The German commander reported tartly that the capture of this outpost was not even noticed by other German troops. The Germans changed their tactics but could not stop Australian infiltration. German troops dreaded the sectors opposite the Australians, and it was at this time that captured

German maps showed the Australian positions marked 'Shock troops'. Later it was found that both Hindenburg and Ludendorff had referred to Australians as shock troops as early as 1916.

There is at least one recorded instance of a German unit refusing to be put in the line opposite the Australians. 'These troops do not fight according to any recognised conception of combat,' the letter of a German officer said plaintively.

For once, British-controlled troops were using their initiative and fighting according to opportunity and local knowledge.

It is not too much to say that the Australian successes with peaceful penetration greatly raised Allied morale and showed that the Germans were in no way invincible. They had no answer to Australian tactics.

Some of the encounters were bloody enough. On 10 June the 7th Brigade, under Monash, seized the latest German trench system at Morlancourt, capturing 325 prisoners and inflicting 400 casualties. This feat must have sent a chill through the German High Command. The German commander on the field reported to his chiefs that a complete battalion had been wiped out as with a sponge. 'Unless steps are taken to prevent it there could be a repetition of this occurrence on a larger scale,' he warned.

On 3 July, the 'occurrence on a larger scale' happened. With brilliant planning by Monash, Australians, supported by tanks, made great gains at Hamel. The Australians had been let down by the tanks at Bullecourt, but now they were much improved and co-operation between Australian infantry and British tanks was splendid. The tank unit commander commented on the superb morale of the Australian troops, who 'never considered that the presence of tanks exonerated them from fighting, and took instant advantage of any opportunity made by the tanks'.

The Hamel success resulted in a personal visit from 'Tiger' Clemenceau and from that attack until the end of the war, Monash's planning was the model for almost all the infantry tank actions.

The Times History notes that at Hamel 'the Australians won even more than their usual distinction'. This first victory in the spring campaign suggested by its ease and completeness that the Australians were capable of taking the offensive on a great scale, and defeating the Germans. Other battles up to now had been won by sheer hard fighting; the losses on both sides had been approximately equal. This was the first battle of the last year, in which the Australians did exactly what they wanted, and everything went according to programme, the first forecast of the overwhelming victories that were to come in the late summer.

'It was by the lessons learnt at Hamel,' General Rawlinson said, 'that they were able to organise and carry through the extraordinarily successful offensive of 8 August.' He said that this was the only instance he remembered in the war when a corps which had been allotted certain difficult and highly important objectives was able to carry out a complete success by winning those objectives, exactly as previously arranged, and half an hour before the scheduled time.

Peaceful penetration prevented the Germans from strengthening their front. Numerous orders came from the German High Command. General von Marwitz, German Second Army, wrote: 'Troops must fight. They must not give way at every opportunity and seek to avoid fighting; otherwise they will get the feeling that the enemy is superior to them.'

The Australians *were* superior—and they knew it. Confidence like this could hardly fail.

On 29 July, shortly before leaving the line, the 5th Australian Division took over a large slab of the German defences at Morlancourt. The bag included 128 Germans and thirty-six machine-guns. The enemy was so infuriated he brought up a fresh division to make an attack. When this was launched on 6 August the Australians had been relieved by a British corps which was pushed back, losing 282 prisoners.

Perhaps the best tribute paid to the Australians' peaceful penetration was that by a British officer who wrote in the British Fourth Army's war diary that the Australian corps had gained a 'mastery over the enemy such as has probably not been gained by our troops in any previous period of the war'.

The Diggers were well to the fore in the great British attack on 8 August 1918. By 7.30 in the morning, the 2nd and 3rd Divisions overran and captured the German field artillery and dug in while the 4th and 5th Divisions moved up to take over the attack at 8.20. To the south were the Canadians; the Diggers always liked working with Canadians.

The 4th Division rushed the valley at Morlancourt, capturing many German support and reserve troops, and by the end of the day it was clear that the Australians had played a major part in one of the great British victories of the war. In his book, *My War Memories*, Ludendorff called 8 August the black day of the German Army.

British cavalry, tanks and planes and French infantry had all played their part in this offensive, in which 16,500 prisoners were taken.

The 1st Division arrived at the battlefield on the 8th and was in action on the 9th. By sheer fierce infantry fighting the Australians and Canadians made an advance of nineteen to twenty-three kilometres. Total Australian casualties amounted to 6000. Then, for a week after the main attacks, peaceful penetration went on along the Australian front.

Following this, the Diggers took part in several major attacks. On 23 August, attacking Herleville and Chuignes, the 1st Division captured 2000 prisoners, but itself suffered 1000 casualties.

Two incidents which took place during this attack later tended to overshadow the attack itself. One of these incidents was probably the most effective single feat of peaceful penetration. The prime actor in the piece was Lieutenant L .D. McCarthy, whose company was supporting the British flank. The end British battalion was pushed back, whereupon McCarthy and his platoon sergeant, and later two English soldiers, attacked from the flank. The tiny party killed a number of Germans, took forty prisoners and handed over to the British 640 metres of captured trench.

According to a story I heard from a member of McCarthy's company, the British officer said: 'It's absolutely ridiculous, old boy, but thank you.' McCarthy was thanked also with the award of a VC.

Possibly the most remarkable feat of initiative occurred on 9 August 1918, when the British made an advance to capture Chipilly Spur. Strictly speaking, the attack was no business of the Australians, but from their positions they could see that A Company of the 2/10th Londons next to the Somme were sheltering about 800 metres from Chipilly, and were, apparently, unable to go forward. Major Mackenzie of 1st Battalion sent a patrol of two NCOs and four men across the river to find out what was holding up the attack and if possible to help it forward.

The patrol leader was CQMS Hayes, DCM, of New South Wales, and the other NCO, Sergeant (later Lieutenant) H. D. Andrews, DCM, of Wauchope, New South Wales.

Crossing the river at six o'clock in the evening, Hayes reported to Captain Berrell of the 2/10th Londons. Berrell said that fire had pinned his company down and advised the Australians not to go forward, but nevertheless the six Australians made a dash for it and got through. Following this example Berrell brought up half his company. The Diggers searched the village of Chipilly in two parties, worked up the spur northwards and found a party of Germans intent on their front. Two privates were left to guard the entrance to the village and another was sent to the 2/10th Londons to guide up a Lewis-gun crew.

Sergeant Andrews with the fourth man pushed around the back of the spur. Hayes, after watching two German machine-guns firing in a valley, led a platoon of the British to a chalk-pit, but a heavy British smoke barrage drove them back. Andrews now took Hayes along the track he had already scouted, almost a kilometre around the reverse side of the spur, and there sighted a small German post. While Andrews and the private covered the post with fire Hayes worked to the flank. As he rose to rush the post he found himself looking into another post of three men. One fired, but missed, then Hayes shot him. The other two Germans bolted, but Andrews and the private captured them.

Telling the 2/10th Londons to follow closely, Hayes, Andrews and two Australian privates returned under the smoke to the post previously attacked. The Germans here retired, but the four Diggers found a stronger post and rushed it, bagging with bomb and bayonet an officer and thirty-one men. The privates, Kane and Fuller, went on and captured nine more men and two machine-guns.

Sergeant Andrews set up a machine-gun and fired on Germans retiring from the ridge. With Fuller and Kane he took another thirty prisoners.

At ten o'clock, having led the British advance all the way, the patrol went home to its own unit. CQMS Hayes had the foresight to ask Captain Berrell for something in writing and Berrell gave him a note recommending the Australians for 'their conspicuous work and magnificent bravery with me today'.

Unfortunately, the incident was not mentioned in the official records of the 2/10th Londons. To all intents and purposes the six Australians captured Chipilly Spur and handed it over to the British. But for Captain Berrell's note and later admissions, all knowledge of Australian help could have been denied.

13

Climax at Mont Saint-Quentin

Many times in 1918 most Australian leaders felt that the Allied High Command did not give the Australians enough scope for thrusting through the German defences, but circumstances presented Monash with a chance to show what Australian troops could really do when working to Australian planning.

Near Péronne the Australian 2nd Division rushed the Germans from a strategic river bend on the Somme and occupied the whole of the western bank. North of the river the 3rd Division had made many gains, capturing Bray, Susanne and Curlu. After dark on 29 August the 38th Battalion, after eighty-nine hours of continuous effort, reached the eastern end of Cléry village.

It now seemed to Monash that if he could transfer his main strength to the northern side of the Somme he could surprise the Germans and rush Mont Saint-Quentin, the key to all strategy in the region, since it overlooked many kilometres of country, including Péronne.

General Rawlinson, the British commander, thought Monash had little chance of bringing off the feat, but gave him permission to try. Australian engineers, sleepless yet working like dynamos, built several bridges across the Somme. Then various Australian infantry units, by hard fighting and skilful manoeuvring, set the stage for the attack on Mont Saint-Quentin by the 5th Brigade of the 2nd Division.

The 20th Battalion had to clear enemy trenches for 1½ kilometres in depth, using grenades and rifles alone, to reach the starting-point for the assault on 31 August. To do this they captured 120 Germans and eleven machine-guns.

The confidence of the Australians was high and at this time they were psychologically unbeatable, though they would not have thought this of themselves. They were ripe for combat and victory. Although the usual Australian practice was to issue rum after action, on this occasion it was issued before action—partly because the troops had been moving ceaselessly for two days and nights.

The attack was, technically, a brigade stunt, but in fact only 600 men made the initial attack—that was the number still left in the 17th and 20th Battalions. The brigade's two other battalions followed in close support and reserve.

The attack began at 5 a.m. on 31 August, with Australian field artillery softening the approaches. Then the charge began, the Australians yelling like demons to give the impression of greater numbers. They routed the forward Germans, men of the 2nd Guard Division, one of the best units in the German Army. A German regimental history said later: 'It all happened like lightning and before we could fire a shot we were taken unawares.'

As crowds of prisoners were filtered to the rear by the support battalions, the charge went on, the Germans crowding the face of the mount and fleeing before the Australian bayonets. The Diggers swept up the slope and over the summit, flushing out the German support troops. For the Germans the position was one of the utmost confusion and at no time was any commander able to stem the rout. A few Germans escaped, but most were rounded up on the mount. General Rawlinson heard about the capture of the mount as he shaved before breakfast. He was a very surprised man and Monash a very pleased one.

The 5th Brigade was so much below strength that it could not keep its full gains. Part of the German 2nd Guard Division, in reserve, drove back the scattered troops from Mont Saint-Quentin village on the crest. 'Even good Australian troops were by no means invincible if strongly attacked,' the German history said. This was a high compliment.

Nevertheless the Diggers held on just below the summit of the mount. And by 7 September units of the 2nd, 3rd and 5th Divisions took Péronne and made other gains, almost crippling five German divisions. The whole action was brilliantly conceived and executed and with other attacks by the

British Third Army and the Canadians, forced the Germans
to retire to the Hindenburg Line.

> The capture of Mont Saint-Quentin, the citadel of Péronne
> [*The Times History* records], was more remarkable than
> Hamel, a combination of valour and skill. General Raw-
> linson has described it as a Gibraltar, commanding the
> passages of the Somme and the access to Péronne. So strong
> was the position that he could not bring himself to order
> troops to attack it, and the suggestion that they should be
> allowed to make the attempt came from the Australians
> themselves. The German commander of Péronne, who
> was captured in the fighting, expressed his admiration of
> the feat. He had believed his position, which was held by
> picked volunteer troops, to be absolutely impregnable.

On 18 September the Australians did even more than their
own commander, Monash, believed they could do—and *his*
confidence in them was great enough. The Australian units
were so very much below strength that Monash merely asked
battalion commanders to make an honest attempt to take the
further 1½ kilometres of country asked of them in the overall
British plan.

The 1st Division captured not only two, but three objectives, and the 4th Division worked its way systematically to
the Hindenburg outpost line. On the southern flank the two
attacking companies of 46th Battalion—just 160 men—got
through enemy wire and took their third objective, capturing
550 Germans and routing many more. By dawn on 19
September most of the Australian line overlooked Saint-
Quentin Canal and the Hindenburg Line.

A glance at battle figures is interesting. The British attack
involved ten divisions in all and between them they had
taken 12,000 prisoners and 100 guns. But of this total the
two Australian divisions captured 4300 prisoners and seventy-
six guns.

Several captured German officers said that their men would
no longer face Australian troops. 'You cannot fight men who
know they cannot be beaten,' a German colonel said despair-
ingly.

Sister N. A. Pike, the author's mother, with convalescent soldiers at the Australian military hospital in Harefield, Middlesex, England, in summer 1917.

Diggers just before passing through their own wire on patrol at Tobruk in 1941. (*Imperial War Museum*)

Captain Albert
Jacka, VC, MC and
Bar, 14th Battalion,
AIF. (*Australian
War Memorial*)

Sergeant Tom
Derrick, VC, DCM,
2/48th Battalion.
(*Australian War
Memorial*)

The Australian victories were even more notable than most people realised, because the troops had been worked and fought to the limit of human endurance since March. No battalion could put more than 300 men in the line and this led to the decision to disband a battalion in each of eight brigades to reinforce sister battalions.

What happened then was remarkable even for the AIF. The men refused to disband. The officers left, but the NCOs and men carried on as if nothing had happened. Discipline was strict and nothing was done to antagonise senior officers. All they wanted to do, the men said, was to go into the next fight in their old units. Not until two weeks later, when the divisions were at rest, was the disbandment carried out. No disciplinary action was taken against the troops.

The next fight was the last for the first AIF. Reinforced by American troops, the two Australian divisions took part in the attack on the Hindenburg Line late in September, and in places broke right through the three systems of the Hindenburg Line. The last action was a brilliant one—the capture of Montbrehain by the 6th Brigade, again to Australian planning.

Two incidents during the battle of Montbrehain were particularly notable.

By skilful use of ground, a party of the 6th Machine-Gun Company under Lieutenant N. F. Wilkinson of Kew, Victoria, worked to the flank of a row of German machine-guns. Wilkinson ordered two belts from each of his two guns. The resulting fire, followed by the machine-gunners pushing on in person, put fourteen machine-guns out of action, killed thirty Germans and captured fifty.

The leading figure in the second exploit was Lieutenant (later Captain) G. M. Ingram, VC, MM, of the 24th Battalion. Aged nineteen and only recently commissioned, Ingram led a charge in which forty Germans were killed and six machine-guns captured. He rushed and captured several small posts, then jumped into a quarry, where he shot several more Germans and captured many more; sixty-three surrendered from one dugout. Also, forty machine-guns were found in the quarry. Ingram, by this time apparently intent on wiping out the German Army single-handed, crept up to a house, shot a gunner firing from it, then burst into the house and captured another large party of Germans. At this stage the opposition ran out.

During the dark days of 1918 a British officer wrote: 'May we be quickly relieved and may the relieving troops be Australians.'

British officers and troops welcomed the Diggers for their abounding willingness and virility. A British colonel wrote in his diary:

> To have Australians near them acted like a tonic on my men. The Australians were always so bright and cheerful; I never once saw them downhearted. They would come into the line with an extraordinary aggressiveness that never failed to hearten other troops about them. The presence of Australians on our flanks always gave us about fifty per cent extra confidence.

In 1918 General Mullens of the British 1st Cavalry Division wrote to Monash: 'As you know we had a curious collection to deal with. It was a very great relief to know I had your stout-hearted fellows on my left flank. . . .It was a pleasure and an honour to be fighting alongside troops who displayed such magnificent morale.'

Monash himself had no illusions about his troops. His maxim was 'Feed your troops on victory'. He wrote in his book, *Australian Victories*: 'They [the Diggers] learned to believe, because of success heaped upon success, that they were invincible. They were right and I believe I was right in shaping a course which would give them the opportunity of proving it.'

Monash believed that the Australian's astonishing success as a soldier was due to several factors, including the democratic institutions under which he was reared, his advanced system of education, teaching him to think for himself and to act decisively and practically, his instinct for sport and adventure and his pride in his young country. 'The Australian Army is proof that individualism is the best and not the worst foundation upon which to build collective discipline,' he wrote.

A German correspondent, after Dernancourt, said that the Australians and Canadians were the best troops of the British forces. A sergeant-major of the German Tank Corps, during interrogation after his capture, said that it was generally considered that the Australian troops were the finest in the world and that the Germans were loath to attack them.

Not all German commanders considered the Australians their toughest opponents. General von Kuhl appears to have considered the Canadians the best troops, and Crown Prince Rupprecht several times referred to the fighting qualities of the Scots.

It is no coincidence that throughout the whole of the British defensive the Canadian Corps was maintained in position for counter-attacking if the Germans should capture Vimy or Arras. Nor was it a coincidence that the Australian 1st Division was sent to guard Hazebrouck, on Haig's personal insistence. Haig wrote to Birdwood on 15 April 1918: 'The right of our line in close connexion with the French is so vital to me that I must keep reliable troops there and I cannot tell you with what confidence I contemplate the situation in that part of my front as long as the Australian Corps is holding it.'

Captain G. H. (later Sir Hubert) Wilkins was standing by the roadside one day watching French infantry arrive with their ramshackle transport to support General Plumer's British forces. Two Frenchwomen standing behind him tapped Wilkins on the arm and one said: 'Français soldiers good soldiers, like the Australians. Not much salute, march all over the road, officers talk with the men like Australians, but good soldiers.'

The most sincere opinions came from the Germans and some have already been listed. Not all opinions were complimentary. One night in France the 48th Battalion caught a German. He was asked if the Germans knew who was in the line in front of them. 'Oh yes,' he said. 'We always know when the Australians are in the line by the way they walk about the top of the trenches.'

The 50th Prussian Reserve Division apparently had a grudge against the Australians, because they showed more brutality to prisoners than any other German unit. After the capture of seven survivors of the Battle of Dernancourt a German officer asked who they were. Private F. Curtis said: 'Australians.' The officer immediately drew his pistol and shot Curtis through the stomach. The Australian died soon after his mates carried him to the rear.

On another occasion Private V. Savage, a Queenslander of the 47th Battalion, was captured after being wounded in both elbows. On his way to the German rear he asked for a drink. The Germans asked who he was and when Savage said he was an Australian they hit him in the mouth and threatened to bayonet him if he made a sound.

Still the Germans were not always brutal. After one fight a German doctor in charge of a field hospital treated the wounded alternately—one Australian, then one German.

It would be going too far to say that the Australians stopped the great German offensive, but it is true that they saved Amiens, one of the major pivots on the British front.

From 25 March 1918, until 5 October 1918, the day they came out of the line for the last time, the Australians captured 29,144 prisoners. Between 8 August and 5 October they recaptured and released 116 towns and villages, apart from hamlets, farms, factories and other installations.

At this time the Australian Corps represented only 9½ per cent of the total British Army on the Western Front, but its capture in prisoners between 27 March and 5 October was 23 per cent of the total, in guns 23½ per cent and in territory reoccupied 21½ per cent.

In praising the infantry it is easy to omit the artillery, pioneers and machine-gunners, among others. Often enough pioneers fought as infantry and very frequently they were subjected to intense bombardment. They suffered terribly on the nights of 2–3 August 1916, while digging jumping-off trenches for an infantry attack. Men were buried as they worked, dug out and buried again. In ten days the 4th Pioneers lost eight officers and 222 men, mainly in keeping open one section of a communication trench.

Pioneers, because of their very duties, are tough troops, as they showed in World War II when they fought as infantry in Syria.

Duckboards were probably the invention of the Pioneers, though nobody now knows the name of the man who first thought of them. Slatted boards nailed to long runners, duckboards were used as short bridges and kilometres of them covered the floors of trenches and wound from base lines to the front-lines, making progress through the mud much easier.

Australian artillery also took a lot of punishment in the Great War, and they often fought for days and nights on end. Whenever they were supporting their own men they knew no rest. The liaison between infantry and artillery was remarkable. The Australian artillery drivers were noted all along the front for the way they took up ammunition through barrages laid on the tracks. Even more than most soldiers they seemed to realise that by pushing straight ahead a job would be better done and the loss lighter.

The first AIF officially ceased to exist on 1 April 1921. The war had cost Australia 215,000 battle casualties—59,242 deaths and 166,818 wounded or gassed. Another 4084 men had been taken prisoner. At the beginning of the war Australia had a population of 4,875,325, yet the country raised a force of 416,809. Of these 331,781 took the field. No fewer than sixty-five men in every 100 became casualties.

What sort of dividend did Australia get for such a tremendous cost? Materially, very little; certainly not enough to offset even one Australian life. But it did win a startling reputation, which Australians have never fully realised. The Diggers had impressed the world; they had given Australia an international reputation and left a legacy of courage which one day could save its life.

14

The Light Horse—36 Battles, 36 Victories

Early in 1916 the outlook for the British in the Middle East was poor. The Turks already held much of north and west Arabia and were inflicting another major reverse on the British—at Kut, Iraq, where an entire army was trapped and about to surrender.

At stake was the Suez Canal. Its loss would be a major disaster—warships, troopships and supply ships would be forced to make the long journey around the Cape of Good Hope. Political consequences would also be nasty. News of Allied defeats on the Western Front of France and Belgium was already causing doubts in Arabia, India and Africa over British ability to defeat Germany–Austria and their Turkish allies.

At this time of crisis, in March 1916, the Anzac Mounted Division was formed in Egypt and entrusted to the command of an unknown Australian Regular Army major-general, Harry Chauvel. Wiry and tough, alert and intelligent, Chauvel became in thirty months the greatest leader of horse soldiers in modern times. His troopers achieved results un-equalled by any other division of horse, Allied or enemy, on any front of the war in World War I. Between 1916 and 1918 the Australian Light Horse and Camel Corps fought thirty-six battles between the Suez Canal and Damascus—a distance of 640 kilometres—and won all of them.

In 1916 all this was in the future but even at that early date British commanders in the Middle East were worried that the War Office might order the Anzac mounted troops to France. As the infantry left Egypt for Europe, General Murray cabled: 'I am assuming that you are leaving the three

Australian mounted brigades and the New Zealand mounted brigade with me. Otherwise I shall be deprived of the only really reliable mounted troops I have.'

After the heavy casualties of Pozières Birdwood asked for Light Horsemen to reinforce the Diggers in France as infantry. Murray strenuously objected: 'I cannot spare a single man for these reinforcements. These Anzac troops are the keystone of the defence of Egypt.'

The defeat of the Turks at Romani in August 1916 and at El Arish in December 1916 was the prelude to an unbroken and victorious campaign for the Light Horse and Camel Corps. In the final offensive of September 1918 General Allenby used the Australian cavalry and Camel Corps as his spearhead. Swift and terribly efficient the Australians took Damascus from the Turks who had held it for four centuries.

The squadrons of Harry Chauvel were victorious where even Napoleon had failed. The whole campaign was fought in great heat and discomfort and the supplying of a large mounted army fighting in such difficult country was a triumph of skilful organisation. Murray wrote: 'Any work entrusted to these excellent troops is invariably well executed.'

Members of the Light Horse were quick to show their natural inventiveness and ingenuity. The troops brought in many unusual methods—the use of spearhead pumps by which horses could be quickly watered, horse-drawn stretchers and sleighs for the wounded, driven rather than ridden wagon teams, sifting and incinerating of all refuse to keep down the flies, even mounted dental units (a New Zealand improvisation) to deal with dental plates and simple troubles. Several of these innovations—pumps, the driving of wagons, dentists—were at first strongly resisted by General Murray's regular army staff. The necessary material was therefore bought from Anzac funds till the staff changed their minds and authorised the changes and in some cases ordered their adoption into the British Army.

As with other parts of the AIF, leadership was paramount, and the AIF never had a better leader than Chauvel, for whom the battle of Romani was a personal triumph which established him as a first-class commander.

Fought on the very days that the 2nd Division in France was capturing and consolidating its hold on Pozières Ridge, the battle of Romani completely changed the outlook of the campaign in the Middle East. In the five days' fighting at Romani the Turks lost half their force; 5250 are believed to have been killed or wounded and nearly 5000 were captured. The British loss between 5 August and 9 August was 1130, all but a few hundred being among the 5000 Anzacs. The task of clearing the difficult Sinai Peninsula had fallen almost entirely on the mounted troops of which four-fifths were Australians.

With the Light Horse it was a voluntary and unwritten law that no sound man should allow himself to be taken prisoner and no wounded man should be permitted to fall into enemy hands. In the 2½ years of their campaigns only seventy-three Light Horsemen and none of their officers were captured by the Turks. In the same time they captured over 40,000 Turks.

Beersheba was another historic battle, for this was the first time that Australian cavalry made a charge—though they had only bayonets and no swords or lances. The outcome of the battle, and for that matter the outcome of the Palestine campaign, depended on this charge, which was just as dramatic as the Charge of the Light Brigade at Balaclava.

It is a peculiarity of English-speaking peoples that they glorify gallant failures and tend to forget equally gallant successes, so that Balaclava, Gallipoli, Mons and Dunkirk, to list a few defeats, are immortal names while spectacular victories—like Beersheba—are forgotten.

The Light Horse charge at Beersheba must have been frighteningly impressive. The horsemen moved off at the trot, five metres between each man, and almost at once they quickened to a gallop. As they came over the top of the ridge they looked down the long gentle slope to Beersheba. Somewhere in front of the town's buildings the Turkish trenches and their garrison lay. The Turkish gunners saw the Australians and opened with shrapnel, but Australians had never ridden any race like this and the pace was too fast for the gunners. After three kilometres Turkish machine-guns opened hotly from the flank but the watchful British batteries at once detected and silenced them.

Next came rifle-fire from the Turkish trenches, dangerous at first, then wild and high as the Light Horse, who could now see the Turks in the trenches, approached. Next the foremost troops were over the front trench and jumping the main one. They dismounted and turned upon the Turks from the rear with rifle and bayonet. The bewildered garrison quickly surrendered. Other Light Horsemen galloped ahead to the rear trenches where parties of fifty Turks surrendered to single men. Other squadrons galloped straight into Beersheba and the day was won. The 4th and 12th Light Horse had only thirty-one killed and thirty-six wounded, and captured over 700 Turks. But above all the eastern flank of the whole Turkish line was turned.

Major C. M. Fetherstonehoare of Coonamble, NSW, had his horse wounded nearly thirty metres from the Turkish trench. With his first revolver shot he put it out of pain and then emptied the magazine into the Turks, capturing a post single-handed.

The last year of the war was the most eventful of all and included several spectacular actions in which junior Light Horse officers proved they they were no less capable and enterprising than their infantry counterparts in France.

In a lightning attack on an enemy camp, a troop under Lieutenant P. W. K. Doig captured 2000 Germans and Turks. In the same general action (by the 3rd Brigade) Lieutenant R. R. W. Patterson, leading a troop, met a large force of retreating troops in the dark. Following a suggestion by a trooper (T. B. George, a Victorian) Patterson opened fire and bluffed 3000 Turks into surrender.

Once again, it was shown that an Australian officer was not too proud to accept a good idea from a private soldier—nor too selfish to give him credit for it after victory. There is a recorded instance in the British Army in Palestine in which a captain had a corporal court-martialled for making a very similar suggestion.

The number of prisoners taken was embarrassing. In nine days the Anzac Mounted Division under General Chaytor took 10,300 prisoners and fifty-seven guns at a cost of 139 casualties.

On 21 September 1918, two Light Horse regiments, the Victorians and Western Australians, about 600 strong, suddenly enveloped the town of Jenin. Galloping at dusk with drawn swords upon the old stone-built hillside town, they were surprised to meet shouting droves of Turks advancing and crying for mercy as they waved white flags. The only resistance was from a detachment of German machinegunners, who were quickly silenced.

The 600 Australians captured 7000 prisoners, including 700 Germans and a substantial cavalry force with 900 horses; also two aerodromes and a huge quantity of war materials including rolling-stock, guns and machine-guns and complete trains of motor and horse transport.

The Germans had fired great dumps of ammunition, petrol and the hangars and workshops on the dromes, but one plane was taken undamaged. Nearby the Diggers found a big cave containing thousands of bottles of champagne and other wines and spirits. Not all of it was disclosed and that night some of the Light Horsemen celebrated with champagne drunk from German helmets.

A detachment of the 2nd Light Horse captured Musa Kiasim Pasha, Commander of the 53rd Turkish Division, with his escort and staff. An English newspaper correspondent with Allenby's forces, described the incident this way:

This officer along with his staff was very calmly driving along in a gharry on the way to his battle headquarters. Suddenly an undreamed-of troop of Australians swooped down. The gharry horse bolted and overturned the vehicle and amid shouts of Australian laughter the general, his gharry and his officers became prisoners, ignominious and absurd. It was a serious loss to the Turks. I saw this unhappy gentleman on his way back, handsome, dignified and infinitely crestfallen. He was upset that the Australians had laughed at him even though they treated him with the dignity due to his rank.

One of the Light Horse's fiercest fights was at Semakh,

where the railway to Damascus touches the southern end of the Sea of Galilee. The Turks, helped by strong detachments of Germans, had built a sort of stockade with engines and other rolling-stock defended by machine-guns. On 24 September 1918, the position was attacked by the 4th Light Horse Brigade (under Brigadier-General Grant) and bitter hand-to-hand fighting developed. In the end, a determined charge by the 11th Light Horse drove many Germans into the sea where they were lanced or drowned. The brigade took 350 prisoners.

Next day they occupied Tiberias, where the fleas gave them more trouble than the Turks. The Arabs say that the king of the fleas holds his court here and most Diggers were prepared to admit it.

Oddly enough, on one occasion during the Palestine campaign Australians fought with the Turks and not against them. This happened on 28 September 1918, twenty-seven kilometres south of Amman, where the commander of a Turkish force 5000 strong told Lieutenant-Colonel D. C. Cameron, commanding the 5th Light Horse, that he would surrender provided the Australians protected his men against the Beni Sakr Arabs. These wild and dangerous men were encircling the Turks and waiting for a chance to move in and kill and mutilate.

Brigadier-General Ryrie moved his 7th Light Horse quickly to the scene and together the Australians and Turks held their positions against the Arabs. The New Zealand Brigade arrived at dawn and the Turks surrendered. Many sick and wounded Turks later died because the Arabs would do nothing to help them. Lieutenant-Colonel T. J. Todd commanded a camp of 16,000 Turkish sick at Kaubub and although very ill himself did all he could to help the Turks. The colonel died at the camp a few months later.

General Allenby frequently visited the Australians, who seemed to fascinate him. 'I never met troops so completely unimpressed by senior rank,' he told a friend in 1918.

Visiting the front-line, he saw Australian Engineers preparing a water supply by digging deep wells. An English war correspondent noted that some of the Australians were stripped to the waist and some were even naked. The general, told that these men had worked for twenty-four hours on end to get a good flow of water, thanked them personally. Allenby was a little off balance as he thanked the nude Australians, who probably only wanted to get on with the job. The English correspondent was concerned about their nudity, especially so near the front-line.

The Diggers nonplussed Allenby more than once. He was inspecting their lines—I believe near Gaza—and came upon a half-stripped Light Horseman seated outside his tent de-lousing his shirt. The Digger glanced at him, but didn't bother to stand up and went on with his work.

Allenby, conscious that some remark was called for, said: 'Ah, picking them out, soldier?'

'Nope,' the Digger said, without looking up, 'just taking them as they come.'

The Light Horse's last action in Palestine was on 2 October 1918, when the 9th Light Horse Regiment outflanked and rushed a Turkish column, capturing 1500. Soon after this eighty-five Turks under a German officer tried to get a machine-gun into action against the Australians. Two Light Horse signallers, one of them a jockey in civilian life, rushed the enemy, grabbed the officer and turned his revolver and the machine-gun against the Turks, capturing all of them. This sort of individual action, made instantly and without waiting for orders, was commonplace in the Light Horse.

After this two-man coup, General Chauvel paraded his battle-stained mounted troops through Damascus, then rife with rioting and looting. The impressive spectacle had its effect—the city quietened down.

A few members of the Light Horse took part in one of the most remarkable military missions in history—that of Dunsterforce to Mesopotamia. Named after Major-General Dunsterville (the original of Kipling's 'Stalky' in *Stalky and Co.*), Dunsterforce was made up of men hand-picked from all the dominion armies.

The AIF contributed twenty officers and twenty NCOs and another two officers and five NCOs from the Light Horse. Ordered to London in January 1918, the AIF men were soon sent to a camp near Baghdad where they had lessons in Persian and Armenian.

General Birdwood had asked for men of 'exactly' the class of Major Harry Murray, most famous fighter in the AIF, and suggested that Murray himself might like to volunteer. Murray did so, but Birdwood then rejected him, apparently because he believed Murray's influence in the AIF was too valuable. For the same reason Albert Jacka and the brothers Captains A. M. and A. S. Maxwell were not permitted to join Dunsterforce.

The Australians acquitted themselves more than favourably in Dunsterforce's activities in Persia and Kurdistan. The diary of one Australian officer records that of a party of eighty-eight officers and 150 NCOs who had to make an extremely arduous march in appalling country, only fourteen completed the journey on foot and entirely without rest: ten of the fourteen were Australians.

In July and August 80,000 Christian Assyrians fled from the Kurds and Turks in one of the worst treks in human history. Day after day, in a column twenty-four kilometres long, they struggled on foot across desolate country trying to escape from the cruel mounted Kurds who followed them.

The Kurds continually raided the rear of the column, killing and wounding and carrying off women and girls for sale as slaves in Turkish harems. An Australian witness to this dreadful march of the Assyrians was Captain Stanley Savige of the AIF. The young captain begged for and received permission to take a small party and hold the Kurds off the retreating Assyrians, while the British cavalry protected the main body on the flanks.

Savige took with him two officers, Captain E. Scott-Olsen, MC, of Newcastle, and Captain R. Nichol of New Zealand, and six sergeants, including an Australian and a New Zealander. Nine men to hold back 1000 Kurds—but what nine men they were! Hand-picked, finely trained, immensely experienced, absolutely fit, this little group had implicit confidence in Savige and in each other.

On 6 August 1918, Savige took his party the twenty-four kilometres to the rear of the retreat, where wounded women and children, abandoned by stronger husbands and fathers, toiled hopelessly along the road. All that evening and the next day Savige fought the Kurds and saved the lives of many Assyrians. He managed to enlist the help of twelve armed refugees, and with them drove the Kurds from a village. Captain Nichol was killed in this encounter.

Day after day, night after night, the little force, by hard riding and fighting, managed to keep the Kurds off. Savige, exhausted, sent a message to the commander of a cavalry squadron asking for reinforcements. The message was intercepted by an English sergeant in charge of twelve Hussars who brought his men at the gallop. Together the men of Dunsterforce and the party of Hussars brought the rear of the column to safety. Because of overstrain and sickness four of the little party were unable to serve again.

The Assyrians left 30,000 of their number by the wayside. Savige was distressed at leaving behind many weak and wounded, but he had no way of carrying them. The expedition within an expedition was a remarkable feat and it was no great wonder that its commander later became Lieutenant-General Sir Stanley Savige, KBE, CB, DSO, MC.

The Light Horsemen of the Great War were exactly the same type of men who had been so successful during the Boer War. Mostly countrymen, they could ride hard and live hard. They were cavaliers of the sands, and the age-old deserts of the Middle East had never seen anything like them. There was no glamour in their life, but in a way they were the glamour boys of the first AIF, men of infinite resource, stamina and courage.

They rode into Australian history when they made that charge at Beersheba; the pity of it is that they did not ride more completely into the Australian consciousness. If ever a force should have captured the popular imagination it was the Light Horse and Camel Corps.

15

'The Greatest Contempt
for Danger'

Australians won sixty-three VCs during the Great War, fourteen of them posthumously.

It has been often said that the Digger, or at least the case-hardened, battle-wise Digger never volunteers. '*Me* volunteer?' I have heard many a Digger say. 'Not on your bloody life, mate!' He means that he won't volunteer for picket duty, or for spud-barbering or for orderly-room running. But he always volunteers if the job is interesting enough.

One case-hardened Digger who liked to volunteer—if the task was dangerous—was Private Patrick Bugden. And he was invariably one of the *first* to volunteer. *The Times* said: 'Bugden's deeds were of the sort specially associated with the many Australians who won the VC.'

Twice he distinguished himself when the Australian advance was held up by strongly defended pillboxes. In the face of devastating fire from machine-guns he led small parties in assaults on these strong-points and silenced the guns with bombs, after which he would capture the garrison at the point of the bayonet. Another time, Bugden saw that a corporal, a mate of his, had been captured and was being taken to the rear. Single-handed, he charged the Germans, shot one, bayoneted the other two and freed his mate. Five times, as a volunteer, he rescued wounded men under intense shell and machine-gun fire, 'constantly showing the greatest contempt for danger'.

Private Bugden volunteered once too often and was killed while on yet another lone mission. His end was inevitable and he must have known it, for no man can take on the enemy and fate so frequently and live, but I fancy that Private Bugden would have said that it was a good way to go out.

How many Australians have heard of Private Patrick Bugden?

Here are some other Great War VC winners and their exploits, selected to reveal Digger characteristics, particularly élan, tenacity and cool, intelligent daring.

Private Thomas Kenny of the 2nd Battalion won his VC at Hermies, France, on 9 April 1917, with a single-handed feat of gallantry. His platoon was held up by an enemy strongpoint with a heavy machine-gun. The Australians suffered serious casualties and no progress was possible. Under heavy fire at close range Kenny ran alone towards the German post, killing one man who tried to stop him. A proficient bomber, he threw grenades into the strongpoint, captured the gun crew, all of whom were wounded, and killed an officer who resisted. The gun was in his hands. When his mates made a bayonet charge they found Kenny in possession of that part of the front. 'Here again,' commented *The Times*, 'was a particularly dashing attack to add to the already brave achievements of the Australians.'

Evacuated with trench feet, Kenny was treated in England and then had some leave. One morning he was riding on a London bus with a friend and, looking up from his newspaper, said, 'Here is the name of my old school friend—he's been killed in action. And in the next column is somebody with the same name as mine who has been awarded the VC.' Near Victoria Station soon after this he was stopped by a military policeman who examined his leave pass and told him to report to AIF Headquarters in nearby Horseferry Road. Here he learnt that the name in the newspaper was his.

Sergeant John Woods Whittle of the 12th Battalion was one of many Australian NCOs to show extreme bravery. By a remarkable charge he had already regained a small trench from which he and his platoon had been forced by the Germans. The enemy broke through the Australian line a second time and Sergeant Whittle's platoon was suffering heavily. The Germans tried to bring up a machine-gun to enfilade the Australian position. Alone, Whittle charged across the bullet-swept ground, killed the entire German gun-crew and brought the gun back to his platoon position.

Whittle was an interesting 'case'. Born in Tasmania in 1883, he had spent a year at the Boer War, five years in the navy and then nine years in the permanent army. He went abroad with the AIF in 1915, became a sergeant and in February 1917 he won the DCM. At one time he was demoted to corporal for 'insubordination'. During a special parade, a colonel had berated Australian troops for rioting when Whittle called out, 'But we are good soldiers though!' Because of his quality as a leader he was later reinstated as a sergeant.

Private Jorgen Christian Jensen of the 50th Battalion rushed a German barricade manned by forty-five Germans, and threw in a bomb. He still had a bomb in one hand and, taking a third bomb from his pocket with his free hand, he drew the pin with his teeth, thus threatening the Germans with two bombs. Jensen told the Germans they were surrounded and they surrendered. Then he sent a prisoner to a neighbouring enemy party with an order to surrender, which they did. The second party was then fired on by other Australians who did not know of the surrender. Jensen stood on the barricade, in great danger of enemy fire, and waved his helmet as a signal for the Australian fire to cease. He then sent both parties of Germans, about seventy of them, back to the Australian lines.

Captain J. E. Newlands of the 12th Battalion won his VC not for any distinctively individual act, but for consistent courage and inflexible tenacity. On three separate occasions he showed extreme devotion to duty when facing heavy odds. First, he organised an attack by his company on a most important objective and under heavy fire personally led a bombing attack. He then rallied his severely punished company and was one of the first to reach the objective. The following night the company was heavily counter-attacked and the captain's personal example and sound judgment dispersed the enemy and regained the position. When the company on Newland's left was overpowered and his own company attacked from the rear, he drove off a combined attack on four occasions.

'It was due to this officer's tenacity and disregard for his own safety that the men held out. His stand was of the greatest importance and had far-reaching results.'

Lieutenant Rupert Vance Moon of the 58th Battalion was ordered to attack a strongpoint near Bullecourt and then a trench behind the strongpoint. Moon was wounded in the initial advance, but reached his first objective. Then he led his men against the trench itself, and was again wounded, this time seriously. Yet he so inspired his men that he captured the enemy trench. He continued to encourage and lead his surviving troops in a general attack, which he successfully pressed home, although he was again wounded. While the position was being consolidated Moon was wounded a fourth time—severely, through the face—and now, finally, in great pain, he agreed to withdraw from the fight.

Captain Rupert Cuthbert Grieve, leading his company of the 37th Battalion, was held up by heavy artillery and machine-gun fire during the attack on Messines on 7 June 1917. All his officers were wounded and his company had seven casualties. Captain Grieve located two machine-guns holding up his advance and, despite continuous fire from both guns, attacked them and killed all members of both crews. He reorganised his company but was wounded, though not before the position was secured.

Private John Carroll of the 33rd Battalion had an amazing fight to win his VC, also at Messines on 7 June 1917. He rushed an enemy trench and bayoneted four Germans. Seeing a mate in trouble he went to help him and killed one of the enemy. Afterwards, coming across a machine-gun and four men in a shell-hole, he attacked the team, killing three men and capturing the gun. Next, under bullet and shellfire, he dug out two of his mates buried by a shell. For the ninety-six hours his battalion was in the line, Carroll showed 'the most wonderful courage and fearlessness'.

Corporal George Julian Howell of the 1st Battalion won his VC with another single-handed exploit, accomplished on his own initiative. During the battle for Bullecourt on 6 May 1917 he saw that a party of the enemy were likely to outflank his battalion, so alone and exposed to very heavy fire, he climbed to the top of the parapet and bombed the enemy back along the trench. When his bombs were exhausted he continued the attack with the bayonet until he was seriously wounded.

'It was a fine spectacle,' said *The Times*, 'and it rejoiced the hearts of the considerable number of men who saw it.'

The infantry had no monopoly of Victoria Crosses, though the very nature of their work gave greater scope for winning decorations than, say, the no less valuable work of the pioneers and sappers.

Sergeant John James Dwyer, of the Machine-Gun Corps, won a VC through the very capable handling of his chosen weapon, a Vickers medium machine-gun. Going forward in the first wave of a brigade assault, Dwyer rushed his gun forward of the captured position so that he could overlook the enemy. When a German machine-gun opened up on the infantry's flank, causing casualties, Dwyer at once moved his Vickers to within thirty metres of the enemy gun, opened point-blank fire and killed the enemy crew. Ignoring the dangerous efforts of several snipers to shoot him, Dwyer carried the captured gun and his own Vickers back to the Australian lines and established both on the brigade's right flank, where he repulsed several counter-attacks.

Next day, Dwyer took several positions under shellfire and eventually his Vickers was blown up. He took his team through the enemy barrage, picked up a reserve gun and returned to action without delay.

I particularly admire the coolness of Private Reginald Roy Inwood of the 10th Battalion who, alone, went forward through a barrage thrown by his own artillery at Polygon Wood in September 1917, and attacked an enemy strong-post, killing several Germans and capturing nine. In the evening he volunteered for a special all-night patrol 550 metres towards the enemy, and returned with valuable in-formation. Soon after this he located a machine-gun causing casualties among the Australians. Again, acting alone and without orders, he went out and bombed the gun and crew. This time he killed all but one German, whom he spared only because he did not want to be encumbered with the captured machine-gun on the way back. It says a lot for

Inwood's coolness and judgement that in the heat of a bloody
and violent action he could have this considered thought.

Lieutenant Valentine Storkey of the 19th Battalion was in
charge of a platoon and on emerging from a wood he encountered the enemy trench line. The lieutenant found himself with only six men, but continued forward. He saw that
about 100 enemy were holding up the advance of troops on
the right and turned to attack the Germans from flank and
rear. He was joined by another officer and four men, and
under Storkey's leadership the little party charged the enemy
with fixed bayonets. The charge was so impetuous and fierce
that the large German party was routed; thirty were killed
or wounded and three officers and fifty men were taken
prisoner. Storkey's exploit took place on 7 April 1918 at Bois
de Hangard, France.

I know the ground on which 23-year-old Captain Clarence
Smith Jeffries won the VC and I frequently stand near the
immensely strong German positions which his battalion, the
34th, had to knock out in the attack on Broodseinde on 12
October 1917. In his citation they are called 'concrete emplacements', in other records they are referred to as 'pillboxes'. Neither term does justice to these powerful
blockhouses of reinforced concrete, impervious to shellfire
and each protected from fire by four other blockhouses.

The 34th attacked uphill through the mud until its centre
was stopped by machine-gun fire from these posts. Jeffries
organised and led the bombing party which rushed and elim-
inated the obstacle; he captured four machine-guns and thirty-
five prisoners. He again led his company forward but the
unit was again harassed by machine-gun fire on the right
flank. Jeffries gathered another party and in between bursts
of fire from the guns got in fairly close. Then he gave the
signal and his men rushed in; the gun swung round, killing
Jeffries immediately. The survivors captured the blockhouse
minutes later.

In tracing the careers of VC winners after the various wars,
it is interesting to note that none of them ever tried to make
capital of his unique distinction, though many of them had
opportunities to do so. At least 50 per cent of VC winners

never even used the letters after their name, though entitled to do so. Officers or privates, they were unassuming both during the war and after it, although two of the First World War VC winners allowed the adulation to go to their heads and became very heavy drinkers.

The attitude of VC winners to their decorations was best summed up for me by Jim Gordon of the 2/31st Battalion, who won his VC in Syria in 1941. A pleasant, uncomplicated West Australian, a private at the time of his award and later a corporal, Gordon said: 'I look at it this way: I was given this medal to hold on behalf of a lot of other blokes. Whatever I did to get it many a Digger also did, but I was lucky—somebody saw me do it. Any man who gets a decoration of any kind is only a representative of the others who deserved decorations.' Any VC winner would agree with this, for while every single VC winner fully deserved his award, hundreds of equally brave men went unrewarded. This was especially so during the Great War when often enough nobody remained alive to say, 'Jack Smith ought to get a VC.'

After the Great War civilians used to comment derogatively about the award of the Military Medal—the junior award for bravery. 'It doesn't mean anything,' these know-alls would sneer. 'The commanding officer would draw lots to see who was to get an MM.'

I believe that in a few units this was so. But what better method of awarding decorations in a battalion whose every member equally deserved recognition for bravery? The troops themselves thought it was the fairest way of making MM awards, although higher awards were never drawn out of the hat.

A soldier of the 5th Battalion, Private Ernest Albert Corey, is the only man in the entire British forces to be awarded four Military Medals. As a stretcher-bearer on the Western Front he certainly deserved his decorations. His mates said that he should have been given a fifth MM just for staying alive. By the very nature of their work the bearers were exposed to enemy fire and were under greater physical strain than most other soldiers.

Lieutenant-Colonel Harry Murray of the 13th Battalion was the most highly decorated soldier in Australian military history, winning the VC, the CMG, DSO, DCM and the Croix de Guerre.

Charles Bean, who knew more about the AIF than any man, wrote in 1932, at the time of Albert Jacka's death, that Jacka should have come out of the war the most decorated man in the AIF.

One does not usually comment on the giving of decorations but this was an instance in which something obviously went wrong [Bean wrote]. Everybody who knows the facts, knows that Jacka earned the Victoria Cross three times. In many cases there may be doubt as to what decorations should be awarded, but there could be no real doubt in these.

During World War II, Australians won nineteen VCs, eight of them posthumously. The comparative figures for the VC in the two great wars are interesting. In 1914–18 British forces (Army, Navy and Air Force) won 475 VCs; between 1939–45 only 106. This proportion applies to Australia and to Canada (sixty-two and thirteen). New Zealand troops won ten Crosses in 1914–18 and nine between 1939–45. The Indian Army won eighteen during the Great War and thirty in 1939–45.

Of the 522 VCs awarded between 1854 and 1914 only fourteen were posthumous. Of the 633 awarded during the Great War 187 were posthumous. But of the 182 granted in 1939–45, eighty-three were posthumous. During the war in Vietnam four Australian soldiers were awarded the Victoria Cross, two of them posthumously. Statistically, it now seems that a soldier's chances of winning the VC and living to receive it are 50–50. Both the VCs awarded to the British Army during the Falklands War of 1982 were posthumous.

In 1975, long after other Commonwealth countries had instituted their own awards for bravery, Australia introduced a set of decorations. They are, in descending order—the Cross of Valour, the Star of Courage, the Bravery Medal and the

Commendation for Brave Conduct. The Cross of Valour is virtually equivalent to the George Cross. The decorations are not exclusively military.

16

Those Who Stayed

In the winter of 1957–58, a few days before I planned to leave England for France, to travel the Somme and Flanders battlefields, I received a telegram postmarked Frou-des-loups, France. The message was brief—PLEASE COME QUICKLY—and it was signed 'Odette Smithson'.

The terse request echoed back more than a year. Then, while driving from Brussels to the South of France, I had stopped briefly at Frou-des-loups, a tiny and very old French village, but stayed long enough to meet Bert Smithson, a Victorian and an old Digger. He had settled in the village after the Great War and bought a farm.

He was one of four old Diggers known to me who stayed behind or went back, yet the other three did not know of his existence and he didn't know of them. Frou-des-loups, about forty kilometres south-east of Mons, was well beyond the Somme area in which the AIF fought.

No French peasant sends a telegram without good reason. I was on my way within hours of getting the telegram. I took the car ferry to Boulogne and drove through ice and falling snow. Northern France isn't a pleasant place in winter—as the Diggers found out in 1917.

I reached Frou-des-loups in a cold, hard dawn—a little cluster of buildings set in a valley with the church spire dominating all. From a distance it looked like a model village set in cotton wool. The French village is the same in all regions. You see no farm-houses scattered about the countryside, the way you might in Australia or England; the farmers all live in the village, huddling together as though for warmth, and they leave it each day to work their fields. Few villages are on a main road, but set on side tracks which go nowhere in particular, and when you see the tiny clusters from the main road they are picturesque and appealing. But life is hard in the villages.

Few people travel farther than the nearest big town. The roads through the villages are of cobblestones and the only shop, usually, is the *boulangerie*. Yet for all the air of hardship these villages have an atmosphere of permanence never found in a large town and still less in a city. The people are close to life and close to death and are unafraid of either. Each village, no matter how small, has its communal cemetery close to the village, and here, for centuries, each family has buried its dead in its respective plot.

When I stopped my car outside Bert Smithson's humble home I could see nobody, but I knew that everybody in the village would know that I had arrived. I knocked on the door and Odette opened it almost instantly and beckoned me in. A buxom, plain woman of about forty-five—much younger than her husband—she kissed me on both cheeks but did not smile, so I knew that my suspicions about the cause of the telegram were correct. I also knew that I was too late.

'I've been driving all night,' I said.

'I know,' she said. She took my hands in hers and they were large, red and capable. 'My husband is dead,' she said, speaking English. 'He died a little after midnight.'

Odette had all the French peasant's calm acceptance of death, so there was no need for me to say much.

'He was very ill,' she said. 'Perhaps this is good, because he had so much pain. I'll get you some coffee.'

I sat by the kitchen fire—used both for heating and cooking—and wondered how Odette could look so fresh and alive when I was so exhausted. 'I would have come sooner had you asked me,' I said.

'I know,' she said, putting a great black kettle on the fire. 'But Bert would not let me send for you. He was not sure that he would die, you see, and he didn't want to make a big fool of himself. You understand?' I understood. After forty years, Bert Smithson the French farmer was still very much an Australian. Odette went on: 'But then, when he knew he would die, he wanted to be with an Australian and you were the only one whose address I knew. And now, to

bring you all this way . . .' I explained that I had been coming
to France in any case, and we sat and drank coffee and ate
dry rolls in silence. Outside I could hear the creaking of
lumbering horse-drawn carts, the sounds of the village beginning its day's work. 'I feel useless,' I said. 'Anything I can
do?'

Odette folded her arms and looked me in the eye. 'Bert
asked that you do something.'

'What is it?' I asked.

She came straight out with it. 'He wanted to be buried in
a military cemetery, with Australians, Allonville.'

I did not know it then, but Allonville is a tiny village with
a population of less than 200, in the Amiens area, and about
150 kilometres from Frou-des-loups. The military cemetery
is an extension of the village communal burial ground. Of
the fifty-five army graves, forty are those of Australians.

Bert's last wish was impossible and I knew it. Commonwealth War Graves Commission regulations forbid burial of
ex-soldiers in cemeteries reserved for those killed on service—
and this is right enough. Odette looked so dismal when I
told her this that I promised to telephone the CWGC re-
gional office in Arras but the response was the one I expected.
They never made exceptions.

When I returned from the next village—Frou-des-loups
had no phone in 1958—the neighbours were calling with
sympathy. Everybody knew Bert's history and they were
flattered, I think, that he had become a naturalised Frenchman—a gesture to his wife, I suspect. Odette was despondent.
A man had a right to his last wish and she wanted Bert's
wish fulfilled. 'You will do *something*, please?' she said, her
brown eyes fixed on me pleadingly.

I said I'd try, but when I left Frou-des-loups and drove to
Allonville I didn't know what I could do. I visited the
cemetery and found that nearly all the Diggers there had
been killed in December 1916–January 1917 (post-Pozières)
or in May 1918. Earlier burials were men killed when a shell
hit a barn in which they were sleeping. The 1918 men died
in the defence of Amiens.

I talked with the schoolteacher, who found the priest, who

brought along the man who cared for the communal cemetery, and within half an hour I was in conference with the
adult population of Allonville. They made me an offer.
M. Bert Smithson could be buried in the ordinary section of
the cemetery and very close to the military extension which
adjoined it. His body would lie closer to a row of Australians
than the two rows of army graves were to one another. The
Giarud family was prepared to relinquish its rights to the
plot in question. Would this be satisfactory?

I said that Mme Smithson would want to pay for the plot,
but massed hands scorned the idea. I think that Bert's becoming naturalised really brought about the offer, which was a
far more generous one than might appear.

I returned to Frou-des-loups with the news and now Odette
broke down and cried, though more with relief and gratitude
than anything else. Perhaps, she said, she would go to live
in Allonville, but of course she didn't; her roots were far too
deep in Frou-des-loups. Tired out, I went to sleep with the
church bell tolling evening meditation. Snow was falling
again.

We set off for Allonville next morning. I took the eight
people who could spare the time, and a makeshift hearse
brought Bert's body. The burial took place in mid-afternoon.
Odette didn't cry, but I did. Not because of Bert Smithson;
I hardly knew him. But because every individual in that
village came to the cemetery. It was a gesture that meant
much. We stood ankle-deep in snow during the brief service
and when it was over the villagers spoke to Odette and to
me, for they were under the impression that I was a relative;
by now I was beginning to feel like one.

Then the children of the village put a few flowers on each
of the fifty-five soldiers' graves. Not bouquets, wreaths or
posies, just a few loose flowers gathered from only the children knew where in that great expanse of snow. The whole
populace moved about among the graves, so neat, so uniform,
so upright and soldierly.

Eventually, we left the cemetery—Odette and I leaving
last of all and our footprints in the snow were not all that
we left behind in that oddly peaceful place. I couldn't help

thinking that at long last Bert Smithson had caught up with his mates. They had not grown old as he, who had been left, had grown old, but now it didn't matter.

It was quite dark when we returned to Frou-des-loups and soon Odette's kitchen was full of people asking about the service. Was it a good cemetery? Could you hear what the priest said? And now, suddenly, seeing this new widow among her own people I felt superfluous . . . an intruder. I knew I was feeling more emotion than they did and I didn't want to show it, not any more than Bert Smithson wanted to appear a big fool.

I offered to help Odette in any way, but she said she could manage and naturally she did. There were few loose ends to Bert's life, because he had travelled light. His kin were all long dead and any papers he had he destroyed before getting out of the village at the beginning of the German Occupation in 1940.

When I left Frou-des-loups next morning it was still snow-covered, just as Allonville would be, like something from a postcard. In a few hours the winter's day would end and dusk would fall, the countryside would be bleak and nothing would move on the low hills and shallow valleys. But somehow it was not depressing and though my fingers and feet were numb with cold as I drove away, inwardly I felt as warm as the hearts of the French peasants had been towards the Diggers of 1916–18.

On the wall of every classroom in the public school of Villers-Bretonneux is a placard with the words N'OUBLIONS JAMAIS L'AUSTRALIA—Never Forget Australia. This motto, still so conspicuous many years after the last Digger left the Somme, is a fair indication of the regard in which the French people hold the Australian soldier.

The people of Villers-Bretonneux, where the AIF stopped the German advance in April and May 1918, and so saved the city of Amiens, have a special reason to be grateful to Australia. After the war Australian money rebuilt the devastated town and the schoolchildren of Victoria donated the money which raised the new school. A lengthy inscription on the school's foundation stone bears witness to this.

When I first visited the school a class of seven-year-olds chanted me a song about the 'jumping kangaroo'. I suspect that it was composed by a Frenchman because the kangaroo described in the song was a queer animal indeed, but again it was a moving indication of the town's awareness of Australia. One of the main streets is Rue de Melbourne and one of the main shops is A la Ville de Melbourne.

A few kilometres out of town stands the Australian Memorial—an imposing monument overlooking the fields, valleys and woods around Villers, Corbie and Hamel. The walls of the memorial are inscribed with the names of 11,000 Diggers with no known graves, while the monument itself faces a cemetery in which lie 772 Australian soldiers.

The monument caretaker at that time was Charlie Atkins, a former Light Horseman who finished up in France with the 2nd Division Artillery. Atkins, from Henley Beach, South Australia, returned to France in 1920 and worked with the War Graves Commission before taking over the monument. He was one of three old Diggers still living in the battlefield areas. Most prominent of the three was Rolly Goddard, owner of the Anzac Hotel, Amiens. Goddard, formerly of the 16th Battalion (Western Australia), was one of the best known men in Amiens. Wounded at Villers-Bretonneux in April 1918, Goddard returned to France at the end of 1919— without ever going home—and opened a mercery-drapery business in Villers-Bretonneux. In 1938 he started in the hotel business, but lost everything during the Occupation. He escorted sixteen British women out of France with the Germans on his heels and later served as a liaison officer with the British Army.

Still very much an Australian and a Digger, Goddard found himself regarded as an unofficial consul, and *gendarmes* and town hall officials called for him whenever an English-speaking visitor was in difficulties. No old Digger wanting to visit the battlefields asked for his help in vain. In 1953 a Digger from Western Australia, touring the battlefields with his wife, was taken seriously ill at Valenciennes, about 240 kilometres from Amiens. The worried woman phoned God-

dard, who abandoned his business and went to her aid, only
to find the Digger dead. It was Goddard who handled all
the unhappy details and saw the widow safely back to England.

Another time a Digger from the 9th Field Ambulance
wanted to return a flag he had souvenired from the fire
station of a village in the Somme. He came to France and
Goddard took him to the village where there was a spectacular parade. The mayor, who was also the fire chief, received
the flag, which had first been owned by his grandfather. 'We
couldn't have been more warmly welcomed if we'd turned
up with the French national debt in cash,' Goddard told me.
'Nobody was critical of the Digger for taking the flag, but I
guess everybody was astonished that he had brought it back.'

After so many years Amiens might have forgotten that
Australians saved the city, but in fact even the children are
taught about it and 'the whole populace is constantly re-
minded of it when they visit the magnificent old cathedral.
On a huge, prominent pillar is a plaque, with the inscription,
in French and English: 'To the glory of God and to the
soldiers of the Australian Imperial Force who valiantly par-
ticipated in the victorious defence of Amiens from March to
August 1918 and gave their lives in the cause of justice,
liberty and humanity this tablet is consecrated by the Government of the Commonwealth.'

The third old Digger representing Australia and the AIF
in 1958 was Harry Waldon, formerly a head gardener with
the War Graves Commission and now retired. Waldon not
only lived in Armentières, but had married a mademoiselle
from Armentières. Aged sixty in 1958, and mother of a
daughter aged thirty-seven, Marie Waldon had all the charm
that captivated the Digger.

A member of the 11th Battalion (Western Australia), Waldon was so unchanged after spending two-thirds of his life
in France that he might have arrived yesterday. I stayed
with him in Armentières and never was a man more generous
or open-hearted—a typical Digger.

Armentières is a big place, but ask anybody in the street
for 'l'Aussie' and they would direct you to Harry Waldon's

modest home. He was gassed in the First War, a prisoner of the Nazis for five years in the Second, but he had no bitterness against the Germans who, he said were worthy opponents on the field.

The old song 'Mademoiselle From Armentières' is still sung in France, but has no reference to the Diggers. The AIF's war-time version went like this:

They laughed and loved in the old French town,
            parley vous,
And her heart shone out of her eyes of brown,
            parley vous,
But time fled by and there came a day
When he and his cobbers were called away,
Inky-pinky parley vous.

Quiet the old estaminet,
            parley vous,
No more the Diggers will come that way,
            parley vous,
May your heart grow light through the passing years,
Oh Mademoiselle from Armentières,
Inky-pinky parley vous.

All of the troops who served in France and Flanders are represented by memorials as are the Australians. Apart from the main Australian Memorial at Villers, each of the five divisions has its own monument. The 1st Division's is at Pozières, the 2nd's at Mont Saint-Quentin, the 3rd at Sailly-le-Sec (near Albert), the 4th's at Bellenglise, near Saint-Quentin, and the 5th's at Polygon Wood, near Ypres, Belgium. The 1st Australian Tunnelling Company has a memorial at Hill 60, near Ypres.

At the small village of Bullecourt in 1980 a monument to the Australians and others who fought there in 1917 was unveiled. It is topped by a bronze slouch hat, eloquent testimony of Digger sacrifice. The small town hall has a museum dedicated to the Australians.

At Vieux Berquin, between Estaires and Strazeele, a monument was unveiled in 1983 to commemorate the courage of the Australians who were thrown in to hold a crumbling front in April 1918 when all seemed lost during the German offensive on the Lys River. It arrived late, this monument, but not too late.

Menin Gate, Ypres, carries the inscribed names of 6000 Australians who have no known graves. Here, every night of the year, the' Last Post' is sounded by Belgian buglers and no matter how cold the night some Ypres citizens attend the ceremony. The 6000 Diggers commemorated on this great arched gate are only a few of the total number of names— 56,000 in all.

There is yet another memorial on Pozières Ridge where three Australian divisions lost 23,000 men in July–September 1916. A motorist could easily miss this memorial, which is a simple bronze plaque set almost flush with the ground. It is on the site of the old windmill at Pozières, and the site, now fenced but left as it was after the battle, is owned by Australia.

The Germans, during the Second War, in no way defaced any Australian memorial, though the main monument at Villers-Bretonneux was damaged by shellfire during the German advance in 1940. The Germans did, however, remove the bronze statue from the base of the 2nd Division Memorial at Mont Saint-Quentin. The statue showed a Digger bayoneting the German eagle.

In 1971 it was replaced with a magnificent outsize Digger, modelled and cast by Stanley Hammond. The slouch-hatted figure, with a steel helmet at his feet but otherwise in full battle gear, is by the side of the road and almost at the crest of the hill which the Australians captured in such style.

Many people played a part in securing a replacement for the first monument but none more so than Pierre Tripet, from Roye. He was only thirteen when the Great War broke out and his father was killed in the 1914 fighting. He was passed from one family to another in the Somme–Arras region and he then attached himself to the AIF's 2nd Division. The 20th Battalion fitted him with a uniform and he became the battalion mascot.

'When I first met the Aussies they fascinated me,' he said. 'It was the slouch hat, the light uniforms, the strong and tough bodies of the men, the soft tread of their march ... marvellous altogether. I did not realise at that time the thick [his word] sacrifice of these valiant men.'

After 1945 he constantly harassed the French and Australian authorities to replace the monument. He was one of the principal guests when a contingent of the 2nd Division visited Mont Saint-Quentin for the unveiling in 1971.

Other less conspicuous references to the Diggers can be found all over the Somme and Flanders. For instance, the central cross of Tyne Cot Cemetery, Passchendaele, is raised on one of the original German strongpoints, while a plaque notes that the ridge was captured by the Australians in October 1917. Tyne Cot, the largest British military cemetery in the world, with 11,781 graves, is the resting-place of 1353 Diggers. Seven-tenths of the graves, 8365 of them, are unnamed—an indication of the ferocity of the fighting around the Ypres salient. Among the Australian VCs buried here are Captain Clarence Smith Jeffries of the 34th Battalion and Sergeant Lewis McGee of the 40th Battalion.

Australians are buried in several hundred cemeteries and churchyards in France and Belgium and I have visited all of them. Some of the smaller cemeteries—really extensions of local village communal cemeteries—are very impressive. A good example is Bonnay, where seventy-five out of the 106 soldiers buried are Australians. The cemetery visitors' book shows that the villagers are frequent visitors to the graves. The Australian Cemetery at Bapaume, where seventy-four Diggers are buried, is often visited by children, who write such comments as 'We are grateful to the Australians', and 'I put some flowers on the graves'.

Civilian cemeteries are depressing places, but army cemeteries have a quite different atmosphere, with their gardens and orderly, unextravagant headstones. Many stones carry no inscription beyond the soldier's name, age, unit and date of death. But all might well bear the words on the headstone of 2486 Private A. Mackay, 57th Battalion, aged nineteen, in Rue du Bois Military Cemetery, Fleurbaix, near Armentières—'This is a finer resting place than even in Westminster Abbey.'

Apart from the large numbers of Diggers in Tyne Cot Cemetery, others with numerous burials are VC Corner Australian Cemetery, Fromelles, with 410, all unidentified; Queant Road Cemetery, Buissy, near Bapaume, 954; Warlencourt, near Bapaume, 461; Pozières British, 690; Serre Road No 2 Cemetery, 699; Heath Cemetery, Harbonnières, near Albert, 958; Villers-Bretonneux 772; Adelaide Cemetery, Villers, 519; Vignacourt, near Amiens, 423; Péronne Communal Cemetery Extension, 512; Étaples, 481; Saint-Sever Cemetery, Rouen, 895; Buttes New British Cemetery, in Polygon Wood, 564; Hooge Crater Cemetery, Zillebeke, near Ypres, 509; and Lijssenthoek Military Cemetery, near Poperinge, 1129.

Even now bodies are still occasionally recovered from farmers' fields, but there is little chance of identifying them.

No middle-aged Frenchman or Belgian living on the Somme or in Flanders is in any doubt about the role played by the Diggers. 'They always used the Australians to hold or take a position beyond the ability of anybody else to hold or take,' a Péronne schoolteacher told me.

What the old French people of the battle areas remember and what the younger generations remember at second and third hand is as an aspect of personality which the French share with the Australians—pride. The French discerned the pride which Captain G. D. Mitchell, MC, DCM, of the 48th Battalion, described in his book *Backs to the Wall*:

> We Diggers were a race apart. Long separation from Australia had cut us completely away from the land of our birth. The longer a man served, the fewer letters he got, the more he was forgotten. Our only home was our unit . . . Pride in ourselves was our sustaining force.

You can see the pride in scores of photographs taken at the time, in the way the Diggers posed for the camera, as a couple of mates in a studio or in a large group at a brigade gas or small-arms school. Feet firmly planted about half a metre apart, sometimes with knuckles on hips, they eyed the photographer levelly. In more casual situations they generally adopted a more larrikin pose, but it says to those who know soldiers, 'Look at me, I'm Australian.'

Australians touring the Somme and Flanders today, and staying long enough to understand how highly regarded the Australians were as soldiers and men, must feel proud. And yet, seeing the serried rows of graves, the thousands of names on memorial walls, they would also feel anger at the great waste. Let it be said that blind British military traditional tactics and inept British leadership sentenced thousands of Diggers to death. Whenever the Australians were allowed to plan and fight their own actions—as in peaceful penetration and at Mont Saint-Quentin—they captured their objectives with few casualties. Whenever the Diggers were used as pawns and shock troops by the British High Command they lost heavily.

17

The Young Diggers

Right from the beginning the second AIF took over the tradition of the original AIF. The new Diggers were intensely conscious of themselves as trustees of the old Diggers' glory and courage. A great many of the new soldiers were the sons of old soldiers with a fine consciousness of how difficult it would be to live up to the standards of their fathers.

There were no lectures about tradition in the British Army fashion, with new soldiers learning about the glories of the regiment which they had joined. The instructors of the young Diggers of the second 4th Battalion or the second 48th did not imbue them with the deeds of the earlier unit a generation before. It was a matter of taking over the entire sentiment of the first AIF rather than the battle honours of a unit. From the moment of enlistment we were immensely proud of being AIF volunteers for service anywhere 'for the duration of the war and twelve months thereafter'; these were the terms of service. Our army numbers had a special significance—with an X added to the first letter of our respective states; thus NX denoted men from New South Wales, WX from Western Australia and so on. Non-AIF soldiers merely had the initial letter, such as N. That X increased the pride we had in ourselves—and the scorn with which most of us regarded those who did not have it. It took a long time for this scorn to dissipate. We were tolerant about *some* soldiers with an N number—those 'Old Diggers' who had joined up again but were rejected for active service abroad. My father was one of this group; he and I were for a time in the same training battalion. It was not unusual for fathers and sons to be serving together in this way.

The Diggers of 1939 were, to some Australians, 'five-bob-a-day murderers'. It was also said that not being able to earn a living any other way some men had joined the army as an easy option. A 1980s writer, from research rather than actual experience, calls them reckless, adventurous, wild and flamboyant. Adventurous and flamboyant can be accepted; reckless and wild are too strong. Certainly they were profoundly proud of being the first to enlist; the 16th Brigade, the first of the new AIF, felt itself to be superior throughout the war.

The second AIF officially came into being on 15 September 1939, when the Australian Government announced its decision to form a new division—the 6th. (The first five had fought in the Great War.) Medical examination of recruits did not begin until early October. On 27 November, the Cabinet decided to send the 6th Division overseas and on 20 January 1940, eleven transports carrying the 16th Brigade and attached troops sailed from Fremantle. They were accompanied by six transports carrying the New Zealand 4th Brigade and were escorted by the British battleship *Ramillies* and the Australian cruisers *Canberra* and *Australia*.

In February 1940 the 16th Brigade was in camp in Palestine, but it was nearly a year before the Diggers saw action. This first action—the capture of the Italian fortress of Bardia—was entirely successful.

In *The Second World War*, Sir Winston Churchill wrote: 'Before Alamein we never had a victory. After it we never had a defeat.' But at the time of Bardia, Alamein was still nearly two years in the future. Churchill did the Diggers an injustice. If he meant that the *English* never had a victory before Alamein, he may have been right. But the Australians had had victories. Bardia was a major victory; so was the capture of Tobruk, the 9th Division's holding of Tobruk and the capture of Syria. Australians played the major part in this last victory.

In any case, Alamein, while a victory for British generalship, was won by Australians and New Zealanders. Churchill himself admitted this when he wrote: 'The Australian 9th Division struck what history may well proclaim to be the decisive blow at Alamein. . . . The magnificent forward drive of the Australians, achieved by ceaseless bitter fighting, had swung the whole battle in our favour.'

Bardia, the beginning of all this, was a tremendous victory and it was just what the new AIF needed at the beginning of its career. It established a precedent for victory; it raised the already high morale of the AIF to an unshakeable level and inspired the desire and habit of victory. The Italians did not know it, but they were fighting, at Bardia, men who could not have been beaten.

Writing home after Bardia, a soldier said: 'Truth is stranger than fiction. Don't smile when I tell you that about 200 Italians, in strongly defended machine-gun nests, surrendered to eight of us when we were investigating a deep gorge at Bardia. They could have wiped us out cold, but were glad to be out of it. We had been taught two Italian sentences—*Mani in alto* (Hold up your hands) and *Apri li mani* (Open your hands), but we did not have to say them, as their hands shot up all right.'

After Bardia Mussolini broadcast that the fortress had fallen to 400,000 'barbarian Australians' and 400 tanks. Bardia fell to a force only a fraction of the size that held it. Unfortunately, the great droves of prisoners and masses of equipment captured gave rise to the impression in Australia that Bardia was a walk-over. In fact, it was a hard-fought battle, savagely contested in many places.

The Diggers captured 40,000 Italians and the bag of guns and equipment included twenty-six coastal defence guns, seven medium guns, 216 field guns, twenty-six heavy anti-aircraft guns, forty-one infantry guns, 146 anti-tank guns, twelve serviceable medium tanks and 115 ineffective tanks and 708 motor vehicles, excluding motor-cycles.

One company of the 2/3rd Battalion took 2000 prisoners at one point, including sixty officers from a single dugout.

The fight began on 2 January 1940, after two weeks of active patrolling. When the leading battalions left the start-line for the fight they were making history and such is the make-up of the Digger that I think most of the men knew it.

The opposition encountered at Post II proved that the capture of Bardia was no bloodless walk-over. Forty-eight men of the 2/6th Battalion, in two platoons, commanded by Lieutenant J. S. Bowen and Sergeant H. B. Gullett (later Major Gullett, MC), attacked the post on 3 January. They cut passages through enemy wire, wiped out a machine-gun post and finally charged towards a trench on the edge of the Italian defences. Heavy fire broke up the charge in eighteen metres and nearly all the Australians were down. A corporal and some men reached the trench on the left, but were killed. Lieutenant Bowen and his sergeant, P. Miller, were also killed—Miller after leading a bayonet attack.

Sergeant Gullett with another sergeant and two privates attacked and silenced several machine-gun posts. Gullett was wounded twice and knocked out, but got back to the outer Italian trench. Only three Australians were now unwounded and Gullett ordered them to leave him and get out, but they insisted on taking Gullett with them. The three unwounded privates held back an Italian attempt to cut them off and eventually they reached safety, though Gullett was hit a third time. They were the only four men to return from the attack that day and the company was led by the company sergeant-major, since all the officers had been killed or wounded.

The entire 2/6th Battalion (Victorians) faced heavy Italian counter-attacks, all of them unsuccessful but very determined. One Australian platoon made a grenade charge in which they killed twenty-five Italians and captured another forty-seven.

Post II, the main strong-point confronting the Victorians— was the last Italian post to surrender—on 5 January. It was then that the survivors of the Australian charge knew what they had been facing. Post II, which would normally be manned by a platoon (about thirty men) held 350 Italians, including twenty-four officers. It was armed with two field guns, several mortars, six anti-tank guns, thirty-nine machine-guns and 350 rifles.

Appropriately enough, one of the bravest men at Bardia was a soldier who had served as an officer with the NZEF in the Great War. He was Sergeant Symington, MC, MM, of the 2/5th Battalion, whose real name was Symington Brown. Symington was seriously wounded in the battalion's advance but crawled in front of his company commander,

Captain C. H. Smith, to protect him and insisted that Smith rest a Bren-gun on his body and fire it. Symington was hit again and killed. Smith himself died of wounds a few weeks later.

The total Australian casualties were 130 killed or died of wounds and 326 wounded. The six battalions of the 16th and 17th Brigades suffered about equally, though the 2/2nd Battalion, with eighty-seven casualties including nine officers, was hardest hit. The 19th Brigade, which was only slightly committed to the battle, lost eleven men.

But Bardia was more than a victory for the Digger. It was a victory for the Australian officers who achieved it. An English general (O'Connor) was in command of the operation as a whole but the attack was planned and executed by Lieutenant-General Mackay of the 6th Division, his chief of staff, Colonel (later Lieutenant-General) Berryman, and his brigade commanders. The success of the battle was entirely due to their skill and thoroughness.

(The 6th Division was supported by the 16th British Infantry Brigade, the 7th Battalion Royal Tank Regiment, a Field Artillery Company, a medium regiment of Corps Artillery and a Machine-Gun Battalion—the Northumberland Fusiliers. The British infantry played very little part in the action.)

Three weeks later Australian staff and Diggers were again victorious—this time at Tobruk, where the 6th Division lost forty-nine killed and 306 wounded. They took 27,000 prisoners, 208 guns and twenty-three medium tanks and 200 trucks.

Then they went on to drive the Italians before them to Benghazi and beyond, apparently invincible, always confident. The accomplishments of the 6th Division during those two months—January and February 1941—were startling enough to be almost incredible. It's a pity that the later deeds of the 9th Division (babes with arms, if not exactly babes in arms, when the 6th Division was fighting) at Alamein and Tobruk tended to obscure the 6th Division's desert campaign.

Thrown into Greece in a campaign that was more political than military, the Digger once again proved that he was.

resilient and game in adversity. He was driven from Greece by the hordes of Germans who invaded from the north, but not before he had fought bitterly to hold the mountain passes.

Greece was yet another instance of Australian (and New Zealand) troops being committed to a fight for purely political reasons. The decision to support Greece was British (largely Churchill's) and the AIF commander, General Sir Thomas Blamey, did not approve of it. The *Official History* notes that in a message to the Australian Government Blamey wrote: 'Military operations extremely hazardous in view of the disparity between opposing forces in numbers and training.'

Nevertheless the 6th Division (17,125 men) went to Greece where 320 were killed, 494 wounded and 2030 taken prisoner. (The New Zealanders suffered about equally.) The 2/6th Battalion (Victorians) were most severely hit, with twenty-eight killed, forty-three wounded and 217 taken prisoner, but all nine infantry battalions, the three field regiments (artillery) and the 2/1st Anti-Tank Regiment suffered heavily.

That many more men were not taken prisoner was due only to the ingenuity of individual soldiers and groups of soldiers who contrived, by one means or another, to escape after the general evacuation had ended. Most of the 2/2nd Battalion (NSW) had been cut off in the Pinos Gorge when the battalion was fighting a rear-guard action, but many of its members worked their way to the coast and eventually reached safety. Only 112 were taken prisoner, remarkably few for a battalion that had been given up for lost.

Many men taken prisoner in the Peloponnese, in southern Greece, escaped from prison camps, from trains and from columns of marching prisoners.

Because of outstanding Australian staff work and intelligent, enterprising fighting by Diggers from battalion commanders down to section commanders, Greece was not the immense débâcle it might have been. The force made a withdrawal of 480 kilometres, along a single road most of the time, fighting one rear-guard action after another. The entire 6th Division could have been engulfed in the German invasion, with extremely serious consequences to Australia.

The evacuation from Athens proved that Australian troops *are* disciplined. Brigadier Savige, quoted in the *Official History*, wrote:

> The behaviour of the troops was outstandingly good and orderly. The column had moved in three heads to the road which led to the Quay. Ranks were maintained and threes were unbroken. Everybody was exhausted but the unit job was always well in hand. Not only were the troops acting splendidly, but there was an atmosphere of complete faith in their officers and NCOs. Their security-mindedness was so great that on the approach of a body of troops I would ask where their officer was. I would make myself known to him and only thus could I obtain the identity of each unit. When they reached the Quay men stood fast in their ranks. Naval officers, who were present at Dunkirk, expressed their surprise at all troops carrying their weapons and equipment and spare boxes of small arms ammunition.

The field regiments were active in Greece and their members proved that the infantryman was no tougher than the artilleryman.

The *Official History* relates that when the 2/2nd Field Regiment was climbing the Brallos Pass, Brigadier Herring, commanding 6th Division Artillery, ordered that two guns be pulled out and sited on the forward slope of the escarpment about two-thirds of the distance up the slope to cover the demolished bridge over the Sperkhios River. They were placed about five metres apart on a mere ledge at the side of the road in the area held by the 2/4th Battalion and with them was an Italian Breda anti-aircraft machine-gun manned by a British crew.

About 6 p.m. on 21 April the first German vehicles emerged from Lamia and began moving south along the straight road on which the guns had been carefully ranged. The gunners opened fire at nearly 10,000 metres and, in three rounds, hit and stopped the leading truck, whereupon the remainder hastily retired to Lamia.

Throughout the night these gunners, and observers perched

about 360 metres farther up the slope, saw the lights of seemingly hundreds of vehicles moving down the pass into Lamia. On the morning of 22 April four enemy guns— evidently mediums—opened fire from a wood south-east of Lamia well beyond the range of the Australians' 25-pounders, and a column of vehicles again began moving south towards the Sperkhios. Each time these vehicles came within range the Australian guns opened fire and as regularly the German medium guns replied, their shells bursting closer and closer until they were landing within about five metres of the gun-pits. One German shell hit a truck carrying smoke shells which exploded and covered the area round the Australian guns with smoke for half an hour. A trailer carrying high-explosive shells was set on fire and the shells began to explode. A dump of charges was hit and exploded, setting the scrub ablaze. Some enemy field guns which had been brought forward to the Sperkhios now began to fire, and the Australian guns replied.

By one o'clock in the afternoon one Australian gun was out of action. At this stage Lieutenant J. R. Anderson (later Captain Anderson, MC), the young officer in command, saw that about twenty trucks had come forward to the foot of the escarpment on the left and were unloading infantrymen there. He and his men lifted the trail of the gun on to the edge of the pit so as to depress it enough to fire down the face of the hill and, using a weak charge lest the recoil should cause the gun to somersault, fired more than fifty rounds into the enemy infantry. This drew heavier shelling from the German medium guns. The Australian crews took shelter, and when they returned found that the carriage of the re-maining gun had been hit and would not operate. It was then 4 p.m. and the duel had lasted eight hours. More than 160 shells had burst round the two exposed guns and dozens of rounds of their own ammunition had been set on fire or exploded and trucks smashed, but not a man on the scarred and smoking ledge had been hit. Anderson sent half his men back up the hill with the sights and breech-block of the damaged gun and with his two sergeants and the remaining crew tried to put the other gun into action. The German guns now opened fire with deadly accuracy. One man was killed and another wounded on the hillside some distance above the guns; then, at the guns, five men were killed and three wounded, one fatally, leaving only eight unwounded, including Anderson. At dusk, after the wounded men had been carried out, Anderson and Gunner Brown, MM, returned to the guns and brought away the sights and striker mechanism, and the discs and paybooks of the dead.

Crete followed Greece and again the Digger was victorious even in defeat. Crete was a crippling blow to the AIF, for three battalions were lost, 2/1st (New South Wales), 2/7th (Victoria) and 2/11th (Western Australia). In all 274 Australians were killed, 507 wounded and 3102 taken prisoner. The New Zealanders suffered heavily, too, having 671 killed, 1455 wounded and 1692 taken prisoner.

Although Crete was taken, the Germans lost many more men than the British forces. Total British killed were 1742; the Germans lost 3986 highly trained fighting men killed or missing. In the confused fighting of Crete, the Diggers, commanded by Brigadier George Vasey (later Major-General), found themselves fighting a number of individual battles, either by company or battalion, as in the case of the 2/1st and 2/11th Battalions who fought it out at Perivolia until forced to surrender. Between them the two battalions lost 120 killed, but had certainly killed 550 Germans and probably twice as many.

One of the oddest military messages in history was sent by Command HQ to this small force. Command HQ wanted to tell the CO to abandon his position and withdraw his force to a port on the south coast for embarkation.

Several messages were dropped by plane, but none reached the CO. However, one message read: 'Waratahs Bulli Puckapunyals St Kilda Gropers Albany Bogin Hopit.'

The message was meant for the New South Welshmen (Waratahs), Victorians (Puckapunyals) and Gropers (West Australians) who comprised the Australian force. Bulli, St Kilda and Albany are all seaside towns south of their respective capitals. 'Bogin hopit' meant 'Get stuck into the enemy; clear out'.

The battle for Crete was one of improvisation. Tactics, weapons, supplies, communication—everything was improvised. The artillery units badly needed guns and General Freyberg, commanding the force, asked that they be sent from Egypt. It is believed that 100 were sent. Freyberg wrote later:

> Sufficient to say that many did not arrive, others came without their instruments, some without ammunition and some of the ammunition without fuses. . . . The gunners . . . were either British Regular Army, Australian or New Zealanders; men of infinite resource and energy; they set to work and one lot made a sighting appliance out of wood and chewing-gum. Another lot of gunners made out charts which enabled them to shoot without sights or instruments.

After Crete a West Australian private, Harry Richards of the 2/11th Battalion, saved fifty-two men from captivity, once again showing that the Australian private soldier can rise to remarkable heights when confronted with a crisis.

Richards found an invasion barge, with 364 litres of petrol, and hid it in a cave. On the night of 1 June 1941, Richards set sail from Crete, with more than sixty men on his barge. German machine-gunners fired at it, but Richards and a New Zealand private named Taylor soon had the barge moving swiftly. They reached the island of Ghavdos, where they refilled the water containers. With food short, Richards appealed for ten volunteers to stand aside and give the others a better chance. Then the others set off again—two officers and fifty men. The barge ran out of petrol and Richards rigged up a sail from blankets. The only food now was margarine and cocoa, which Richards mixed and issued. By 8 June the men were very weak, but Richards knew the Egyptian coast was near, and at 2.30 a.m. on 9 June the barge grounded at Sidi Barrani.

One escapee said later: 'Richards was beyond description; he exercised his care for us in a most masterly manner and inspired everyone of us to keep our spirits up.'

Another West Australian soldier, Private (later Sergeant)

S. L. Carroll, made a single-handed escape from Crete in a stolen Greek fishing-boat without oars or rowlocks. Although his only food comprised six tins of chocolate and two tins of water, he reached Egypt nine days later, after an incredible feat of makeshift sailing. The experience apparently gave him a taste for the sea for later he transferred to the navy.

18

Syrian Mountains, Libyan Sand

Of all the actions in which Australians have been engaged the Syrian campaign of 1941, fought by the 7th Division, was the least publicised. Beginning when Australians were still escaping from Crete, it lasted only six weeks, but fighting was fierce and unremitting, the terrain difficult and the enemy skilful and determined.

Because of lack of publicity the 7th Division did not receive the credit to which it was entitled. Although certain British and Indian units were used (notably the 5th Indian Brigade Group under Brigadier W. Lloyd) the Syrian campaign was almost entirely an Australian affair. Australian troops involved at the beginning were the 21st and 25th Brigades, 6th Division Cavalry, 9th Division Cavalry, the 2/4th, 2/5th and 2/6th Field Regiments (Artillery), 2/2nd Anti-Tank Regiment; 2/3rd Machine-Gun Battalion, 2/2nd Pioneer Battalion and two 6th Division Battalions, 2/3rd and 2/5th, regrouped after Greece and Crete. These two battalions share the distinction of having fought in more World War II campaigns than any other Australian unit.

The heavy Australian casualties speak for the fierceness of the campaign, fought against the Vichy French, supported by strong Foreign Legion units and black troops. Enemy casualties were severe: 521 killed, 1790 wounded and 3004 as prisoners. About 9000 Vichy troops deserted to the Australian forces. The campaign was a triumph for the Australians, who were fighting a skilful well-led enemy who knew every inch of the country. Soldiers who had fought in the desert, Greece and Crete said that Syria was tougher than any other campaign.

It is difficult to single out from the many successful actions one which is particularly noteworthy, but the attack on Jebel

Mazar by the 2/3rd Battalion was about as fierce an action as any in the campaign.

A combined company of eighty-three men under Captain Ian Hutchinson (later Lieutenant-Colonel Hutchinson, DSO, MC) were to make the attack on Jebel Mazar, a 490-metre spur commanding the Beirut–Damascus road. The already weary company began the climb at 8.30 on the evening of 24 June, but could get no farther than 150 metres from the summit. Exhausted, short of water and depleted, through some sections of the company getting lost at night, the little force was held up by machine-gun and mortar fire. On the evening of 25 June a fresh company of the 2/3rd Battalion under Captain Murchison (later Lieutenant-Colonel Murchison, MC) was sent to replace Hutchinson's weary men and take the summit.

The new company climbed all night and at nine next morning ran into French fire. At seven that night, warned by a sentry that the French were attacking again, Murchison lead an attack against the French, driving them towards Hutchinson's company whose Bren-gunners killed and wounded twenty at a range of almost thirty metres.

At one o'clock on the morning of 27 June Murchison led his men towards the French positions, their approach muffled by a strong wind. Close to the French, Murchison gave orders for attack and a fierce hand-to-hand fight developed. In five minutes six Frenchmen had been killed, mostly bayoneted, and thirty prisoners had been taken.

The climb continued and was so tough that the men had to sling their rifles so as to leave both hands free. About 365 metres from the summit Murchison, now with a platoon only, saw a French machine-gun post silhouetted against the sky. The Diggers quickly overran one gun, but another opened up from thirty metres range. Murchison made for it, firing his pistol, but Private M. V. Melvaine, DCM, charged the gunner at bayonet point and put him to flight.

Murchison led his men another 365 metres to the crest and saw the last French posts only 90 metres ahead. His men were dead tired, but Murchison, with fine, inspiring leadership, urged them into a bayonet charge, shouting at them to

Corporal Jim Gordon, VC, 2/31st Battalion.

Lieut.-General Sir John Lavarack. (*Australian War Memorial*)

Major-General S. G. Savige. (*Australian War Memorial*)

Field-Marshal Sir Thomas Blamey. (*Australian War Memorial*)

yell as they attacked. The French and African defenders, who greatly outnumbered the attackers, panicked and ran.

Jebel Mazar was won. Murchison with the remnants of a company had forced a depleted French battalion off the summit. Small parties of troops who had become detached rejoined Murchison's company. Unfortunately, no artillery support was available and the company came under heavy cross-fire at a range of 900 metres. The French made five attacks that day, 27 June, but each time were driven off. Private C. C. Atkinson (later a corporal) won a DCM for a single-handed counter-attack on the French. Angry because his mate had just been shot through the head, Atkinson took a bag of grenades and rushed the French attackers, breaking up the advance.

At two in the afternoon the French sent an envoy telling Murchison that mortars would blow him off the hill if he didn't surrender. 'If you want to get us off come and do it,' Murchison replied. Yet his position was dangerous. Ammunition was very low and his men were using French machine-guns, rifles and grenades. Food and water were almost gone. There was no sign of artillery support and lower down the hill, Hutchinson's company had been driven off.

A runner got through with a message to hold on at all costs but 'all costs' meant annihilation. Late in the afternoon the French tried to blow Murchison off the cliff. The bombardment caused casualties, but the Diggers drove off an infantry attack. By now Murchison's men had only twelve rounds a man remaining and at least two battalions of French troops surrounded them.

At night, when it was obvious that no reinforcements were coming, Murchison decided to abandon Jebel Mazar and at 10.30 they moved out, carrying their wounded, with Murchison bringing up the rear. They got out without being attacked. During its assault on Jebel Mazar and the subsequent occupation Murchison's company had lost five men killed and eight wounded. Another fifteen were missing, but most of them turned up later.

Brigadier Lomax, commanding the 16th British Brigade and the officer who had ordered the occupation of Jebel

Mazar, wrote: 'Throughout the operation all ranks 2/3 Aust Inf Bn displayed the very highest courage and determination and their dogged endeavour has very justly called forth the unstinted praise and admiration of all ranks of 16 (British) Inf Bde.'

Commenting on an earlier action near Messe, Brigadier Lloyd said: 'The 2/3 Bn acted with the greatest gallantry and dash throughout, the initiative and keenness of the junior leaders being marked. The success . . . in spite of very reduced numbers and fatigue against an enemy in masonry forts . . . was remarkable and worthy of the highest praise.'

Another company commander of the 2/3rd Battalion, Captain P. K. Parbury (later Lieutenant-Colonel Parbury, DSO, MC), engineered a remarkable action on 20 June when ordered to cut the Beirut–Damascus road. After a short and vigorous night action, Parbury succeeded in cutting the road by the simple expedient of placing two telegraph poles across it.

This road-block soon halted French vehicles from both directions. The moment a vehicle pulled up, Parbury's men hustled the Frenchmen away, mostly without trouble. Before long the ninety Australians had eighty-six prisoners from a bag of twenty-six vehicles. Knowing that he could not hold the position after dawn, Parbury arranged for the prisoners to be taken away and he then secured other positions for his company.

The 2/3rd Battalion was not the only one to win distinction in Syria, but its actions were fairly representative of the whole. The 2/16th Battalion at the Litani River and Damour, the 2/25th and 2/31st at Mardjayoun and Jezzine, the 2/27th at Innsariye, all saw a lot of bitter fighting. Even the battalion with the lightest casualties (2/5th) lost two officers and thirty-nine men.

The experience gained by the 7th Division in Syria was very useful a year later when the 21st and 25th Brigades played the major part in stopping the Japanese advance towards Port Moresby, New Guinea.

Cavalry (mechanised) units were always in the van in the Syrian campaign, but probably no single officer saw more action than Lieutenant T. Mills (later Lieutenant-Colonel). Early on 11 June Mills, leading a detachment of tanks and carriers, was held up by Vichy French fire near the Es Sakiye road junction. Mills concealed his tanks and taking with him Sergeant R. T. Cramp, Sergeant R. A. Edwards and Trooper T. D. Killen, he worked his way into a position dominating the French. His little party had a tank-attack rifle, Bren-gun, submachine-gun and rifle. Edwards began the duel with French tanks and anti-tank gunners and eventually the tanks withdrew. Mills and his men killed or dispersed the crew of an anti-tank gun. Mills then went forward, but found a party of French dug in. Making off in another direction, he found himself standing over another trench full of French. His submachine-gun jammed, but Cramp, following up, attacked the first enemy group and both groups then surrendered.

Mills and his party captured forty-five prisoners, two mortars and two anti-tank guns. The interesting thing about the action was that the French were all Foreign Legion men. They weren't so tough after all. For his actions here and elsewhere Mills won the MC.

The Syrian campaign had brought the Digger into conflict with an enemy never before fought by Australians—an improbable enemy, at that, for no Australian would have expected ever to fight the French. The Diggers of World War I had high regard for the French *poilu* as a soldier, but the Digger of this second war had bested him, and had defeated the famed and feared Foreign Légionnaires. The Diggers were outnumbered and they suffered from serious initial disadvantages, not the least of them being that the country was strange to them. That they triumphed over enemy and terrain was yet another instance of Digger adaptability and doggedness.

After the Syrian campaign a French colonel said: 'Until I saw your infantry crossing the Damour River and fighting in the mountains I believed the Foreign Legion were the toughest troops in the world.'

Once again, too, Australian leadership was vindicated. The early stages of the campaign were directed by General Maitland Wilson of the British Army, but active control then passed to General John Lavarack, who showed rather more imagination in his tactics. Also, as he was in closer touch with the fighting he was able to exercise finer judgment than Wilson, who spent most of his time in Palestine.

Two Australian privates accepted, unofficially, the surrender of a small town in Syria. They were Pete McGowan and Max Hickman. Hickman told the story in a letter home:

> The day after the signing of the Armistice we were at an outpost. Everything beyond the occupied area was out of bounds. However, someone said there was a river on the other side of the mountain, and we were badly in need of a wash, so Pete McGowan and I decided to take the risk. In case there were any fish we took a couple of hand grenades.
>
> About two hours' mountaineering brought us into a little township. On the corner as we entered was a chemist's shop, and as we passed the chemist greeted us with '*Bonjour, Messieurs!*' We shook hands, Pete keeping one hand on his grenade (just in case) and by signs the chemist indicated that he wanted us to have a cup of coffee. Meanwhile, he sent a number of men out on messages, and in a few seconds the shop was full, and two or three hundred people had gathered outside. Pete was quite sure we were in a trap, when the crowd parted and a number of fellows whose deportment marked them as notable came in.
>
> The foremost spoke in English, and told us that although Syrian by birth, he was an American citizen. He also told us that Machgarah was a Syrian township, and that eighty per cent of the people were British sympathisers. He assumed that we were Australian officers (as we were the first British troops to enter the town) and had come to see the place before occupying it.
>
> Our Yankee friend introduced us to the Mayor, the town councillors, the priest, and two doctors. They then hurried us away to the police station. Here fourteen gendarmes formed a guard of honour. At the door we were welcomed by the Commissioner and Chief of Police. Presently, accompanied by the Commissioner and the Mayor, we toured the town. The streets were lined with people, who cheered and waved. Then we inspected a number of

houses damaged by our artillery, and assured the owners that they'd get a good hearing when the place had been occupied.

During our tour quite a number of them asked us for permits to visit other towns in the war zone, and we told them it was not safe just then, but arrangements would be made as soon as possible. In the afternoon we were invited to a function welcoming home a prominent citizen, who had been imprisoned for British sympathies at the outbreak of the show.

As we entered the door someone said, '*Les Capitanes*', and everyone stood up until we had taken the seats of honour. We listened to numerous speeches, drank arak and more coffee—and through the interpreter made our good-byes, the Mayor asking that we bring our troops as soon as possible.

It would have been just too bad if an officer had arrived while we were there. As it was, we got our ears chewed a bit when we got back.

Two Diggers won VCs during the Syrian campaign. The first awarded was that to Private Jim Gordon, 2/31st Battalion, whom I mentioned in an earlier chapter. His citation read:

On the night of 10th July, 1941, during an attack on 'Green Hill', north of Jezzine, Private Gordon's company came under intense machine-gun fire and its advance was held up. Movement even by single individuals became almost impossible, one officer and two men being killed and two men being wounded in an effort to advance. The enemy machine-gun position which had brought the two forward platoons to a halt was fortified and completely covered the area occupied by our Forces.

Private Gordon, on his own initiative, crept forward over an area swept by machine-guns and grenade fire and succeeded in approaching close to the post. He then charged it from the front and killed the four machine-gunners with the bayonet. His action completely demoralised the enemy in this sector and the company advanced and took the position.

During the remainder of the action that night and on

the following day, Private Gordon, who has throughout operations shown a high degree of courage, fought with equal gallantry.

The second VC was won by Lieutenant A. R. Cutler, 2/5th Field Regiment (later a diplomatic career officer at ambassador level and then Governor of New South Wales). The citation read:

For most conspicuous and consistent gallantry during the Syrian Campaign, and for outstanding bravery during the bitter fighting at Mardjayoun, when this artillery officer inspired the infantry to press on and his name became a byword amongst the forward troops with which he worked. At Mardjayoun on 19th June, an infantry attack was checked after suffering heavy casualties from enemy counter-attack with tanks.

Enemy machine-gun fire swept the ground, but Lieutenant Cutler pressed a continuation of the attack. With another artillery officer and a small party he pushed on ahead of the infantry and established an outpost in a house. The telephone line was cut. He mended this line under machine-gun fire and returned to the house from which the enemy post and battery were successfully engaged. The enemy then attacked this outpost with infantry and tanks killing Bren-gunners and mortally wounding other officers. Lieutenant Cutler and another manned an anti-tank rifle and a Bren-gun and fought back, driving the enemy infantry away. The infantry continued the attack, but under constant fire from the anti-tank rifle and Bren-gun eventually withdrew.

Lieutenant Cutler then personally supervised the evacuation of wounded members of his party. Undaunted, he pressed a further advance. He had been ordered to establish an outpost from which he could register the only road by which enemy transport could enter the town. With a small party of volunteers he pressed on until finally, with one other, he succeeded in establishing an outpost right in the town, which was occupied by the Foreign Legion. At this time he knew that the enemy was massing on his left for counter-attack and that he was in danger of being cut off. Nevertheless he carried out his task of registering the battery on the road and engaging the enemy post. The enemy counter-attacked with infantry and tanks and he was cut off. He was forced to go to ground, but after dark succeeded in making his way back through enemy lines.

His work of registering the only road by which enemy transport could enter the town was of vital importance and a big factor in the enemy's subsequent retreat. On the night of 23rd–24th June he was in charge of a 25-pounder sent forward to silence an enemy anti-tank gun, and post, which had held up our attack. This he did and next morning the recapture of Mardjayoun was complete. Later at Damour on 6th July, when our forward infantry was pinned to the ground by hostile machine-gun fire, Lieutenant Cutler, regardless of all danger, went to bring a line to his outpost when he was seriously wounded and 26 hours elapsed before it was possible to rescue him. His wounds by this time had become septic, necessitating amputation of his leg. Throughout the campaign this officer's courage was unparalleled and his work was a big factor in the capture of Mardjayoun.

While the 6th Division was in Greece and Crete, most of the 7th in Syria and the 8th in Malaya, the 9th had moved into Tobruk. Seldom in history was it as important to hold a single town or base as it was to hold Tobruk. If it was lost, Rommel could drive on to Egypt, capture the Suez Canal and deprive the British of the entire eastern Mediterranean. Without Tobruk, Rommel was forced to bring all his supplies, ammunition and troops by road 480 kilometres from Benghazi.

Tobruk, Rommel told his Afrika Korps, must be taken. Obviously, the first step was to soften up the natural fortress by bombing. The first raid took place on 9 April 1941. It was only one of 1000 to be made in the next eighty-four days. Once the *Luftwaffe* was in its stride against Tobruk, Rommel threw in heavy tanks, armoured cars and big siege guns. During the first month, when the onslaught was at its fiercest, Rommel ordered attack after attack in an effort to break the perimeter line. The garrison, short of sleep, food and water, stood firm against the ceaseless onslaught. Infantry activity on the perimeter was never-ending. The Australians took 2000 prisoners, both German and Italian, a tremendous bag considering the intensity of the fighting.

During April the AIF won its first VC in the Western Desert campaigns. The award was made to Corporal John Hurst Edmondson of the 2/17th Battalion (NSW) whose posthumous citation read:

On the night of the 13th–14th April, 1941, a party of German infantry broke through the wire defences at Tobruk and established themselves with at least six machine-guns, mortars and two small field pieces. It was decided to attack them with bayonets, and a party consisting of one officer, Corporal Edmondson and five privates, took part in the charge. During the counter-attack Corporal Edmondson was wounded in the neck and stomach, but continued to advance under heavy fire, and killed one enemy with his bayonet. Later his officer had his bayonet in one of the enemy and was grasped about the legs by him, when another attacked him from behind. He called for help, and Corporal Edmondson, who was some yards away, immediately came to his assistance and in spite of his wounds killed both the enemy. This action undoubtedly saved his officer's life.

Shortly after returning from this successful counter-attack, Corporal Edmondson died of wounds. His actions throughout the operations were outstanding for resolution, leadership and conspicuous bravery.

Rommel, whose aide reported him as exclaiming: 'Those damned Australians!' must have been a little anxious by the time his 15th Panzer Division arrived. Confident of this élite force, Rommel dropped the Italians from the side, and made a mighty thrust at the perimeter on 1 May. It was spearheaded by sixty tanks followed by infantry, a special force known as the Panzer Pioneers—combat sappers. It jabbed through the perimeter to a maximum depth of about 2750 metres but the Germans could hold part of their salient for only a few days before the Australians recaptured it in one of the bloodiest sectional fights of the siege. There was no let-up in the fighting or air-raiding, but Rommel seemed to accept the inevitable—that Tobruk could not be taken at that time. In one attack on the north-west sector 200 out of 300 enemy raiders were killed and thirty captured. A South Australian battalion was responsible for this feat.

The Germans were shocked and prisoners made no attempt to hide their feelings. One German medical officer said: 'I cannot understand you Australians. In Poland, France and Belgium once the tanks got through the soldiers took it for granted that they were beaten. But you are like demons. The tanks break through and your infantry still keeps fighting.'

May was another hectic month; never at any time did total quiet descend over the fortress or the perimeter. Conditions were described as 'unbearable' but the Digger is at his best under such conditions. Desert flies swarmed everywhere; the July and August heat was oppressive and exhausting. The water ration of one pint (570 mL) a day per man was hardly enough for both drinking and washing. The staple drink was strong tea, but since it was heavily chlorinated it tasted foul. Still, it was thirst-quenching; one cup and a man couldn't drink more.

In addition to the everlasting heat there were the khamsin winds from the Sahara. These winds picked up every loose particle of sand from the hard desert floor and drove them northwards against the defences. Since infantry action was impossible in these swirling, blinding terrors it was safe for both Australians and Germans to go to ground.

They huddled in their trenches under overcoats and groundsheets, handkerchiefs tied around their eyes and mouths, trying to escape from the myriad minute pellets which, no matter what a man did, found their way to his skin and eyes and tongue. Tank crews closed all hatches and sweated out the khamsins under coats and blankets. Usually, a khamsin would last only half an hour, but it was a very long half-hour. Metal of any kind became unbearably hot during the summer days. Several men lost all the skin off their hands by touching the tops of tanks standing stationary in the sun.

The Germans and Italians were badly disillusioned by the Australian stand. They knew from experience that when a place was besieged the defenders stayed put. They were confounded when the Australians took the initiative, left their defences and made many patrol attacks on Axis positions. Most of these patrol actions were carried out at night and the Axis troops feared them. The Australians moved silently and killed silently, wherever possible. Sometimes the Germans did not know a raid had taken place until next morning, when they found their outpost sections and sentries all dead. It was peaceful penetration of 1918 all over again.

During these months of the Tobruk campaign many classic instances of section, platoon and company actions occurred. According to letters found on one captured German officer the Germans and Italians were in a 'desperate state of nerves and lack of sleep because of the wretched Australians'. But short, informal armistices between the Australians and Germans had not been uncommon during the Great War and they were fairly frequent now. One night the Germans attacked, but were repulsed with heavy loss. Next morning they asked for a truce to collect their dead and wounded and while it was in progress the Diggers climbed out of their trenches to stretch their legs. The Germans did the same and the Diggers waved and shouted to them. 'Heil, Hitler!' 'How would a pint of beer go, mate?' 'Have another go tonight.' And there were other remarks, not altogether complimentary.

Australian stretcher-bearers went out to help dress the German wounded. After the dead and wounded had all been removed a red flare went up to signify that the truce was finished. The Diggers gave a final wave and a 'cooee' and returned to their trenches. Then the war went on again. Truces like this saved many lives. Generally the Germans respected Australian stretcher-bearers and did not fire on them.

Hugh Paterson, a driver in the 20th Brigade, and a nephew of the poet Banjo Paterson, described Tobruk this way:

> *This bloody town's a bloody cuss,*
> *No bloody trams, no bloody bus,*
> *And no one cares for bloody us.*
> *Oh bloody! Bloody! Bloody!*

No bloody sports, no bloody games,
No bloody fun with bloody dames,
This place gives me a bloody pain.
Oh bloody! Bloody! Bloody!

All bloody fleas, no bloody beer,
No bloody booze since we've been here,
And will it come? No bloody fear.
Oh bloody! Bloody! Bloody!

The bloody rumours make me smile,
The bloody wogs are bloody vile,
The bloody Tommies cramp your style.
Oh bloody! Bloody! Bloody!

All bloody dust, no bloody rain,
All bloody fighting since we came,
This army's just a bloody shame.
Oh bloody! Bloody! Bloody!

The bully makes me bloody wild,
I'd nearly eat a bloody child,
The salty water makes me riled.
Oh bloody! Bloody! Bloody!

Air raids all day and bloody night,
Huns striving with all bloody might.
They give us all a bloody fright.
Oh bloody! Bloody! Bloody!

Best bloody place is bloody bed,
With blanket over bloody head,
And then they think you're bloody dead.
Oh bloody! Bloody! Bloody!

The Digger had always found relief in verse like this.

After much experience of Australians the Germans might have been expected to know something of the Digger make-up, but obviously they never learned. They even tried to undermine Digger morale with corny propaganda leaflets, such as this one dropped in its thousands on Tobruk:

> Aussies. After Crete disaster Anzac troops are now being ruthlessly sacrificed by England in Tobruk and Syria. Turkey has concluded pact of friendship with Germany. England will shortly be driven out of the Mediterranean. Offensive from Egypt to relieve you totally smashed. You cannot escape. Our dive-bombers are waiting to sink your transports. Think of the future and your people at home. Come forward. Show white flags and you will be out of danger! Surrender.

The Diggers laughed, made many ribald comments and told the Germans what they could do with their leaflets or what they themselves would do with them.

They held Tobruk and were proud of it. So proud that when Lord Haw-Haw called them 'the rats of Tobruk' they accepted the taunt as a compliment and after the war dubbed themselves the Rats of Tobruk, in the same way that the men of Mons in 1914 were proud to call themselves the Old Contemptibles, although the Kaiser had not intended to compliment them when he referred to Britain's 'contemptible little army'.

The South Africans, who took over Tobruk soon after the Australians left it, lost the fortress to Rommel, a bitter blow to the 9th Division, though they made up for it a year later at Alamein.

But Alamein and the country around it was a Digger battleground for months before Alamein started. A lot of Australian blood was spilled at the Double 24 feature, better known as the Hill of Jesus, during July 1942. In one fight just two battalions—the 2/24th and 2/48th, supported by field artillery and tanks, repulsed extremely heavy German attacks with infantry, tanks and dive-bombers. They not only killed hundreds of Germans, but captured 1150 prisoners— almost one per man to the Australian battalions—knocked out twenty-two tanks and captured much enemy equipment.

On 22 July a Digger of the 2/48th Battalion won a VC at the Hill of Jesus. He was Private Arthur Gurney, whose citation showed that his conduct was worthy of the original 48th Battalion which had fought so magnificently in France. The award was made:

> For gallantry and unselfish bravery in silencing enemy machine-gun posts by bayonet assault at Tel el Eisa, on 22nd July, 1942, thus allowing his company to continue the advance.
>
> During an attack on a strong German position in the early morning of 22nd July, the company to which Private Gurney belonged was held up by intense machine-gun fire from posts less than 100 yards [about 90 metres] ahead, heavy casualties being inflicted on our troops, all the officers being killed or wounded. Grasping the seriousness of the situation and without hesitation, Private Gurney charged the nearest enemy machine-gun post, bayoneted three men and silenced the post. He then continued on to the second post, bayoneted two men and sent out a third as prisoner. At this stage a stick of grenades was thrown at Private Gurney, which knocked him to the ground. He rose again, picked up his rifle and charged a third post, using the bayonet with great vigour. He then disappeared from view and later his body was found in an enemy post. By this single-handed act of gallantry in the face of determined enemy action, Private Gurney enabled his company to press forward to its objective, inflicting heavy losses upon the enemy. The successful outcome of the engagement was almost entirely due to Private Gurney's heroism at the moment when it was needed.

Montgomery's initial attack at Alamein was made by the 9th Division, 51st Highland, 2nd New Zealand and 1st South African Divisions. British armour was used in strength and British infantry divisions made subsidiary attacks but Montgomery knew—as he told me more than twenty years later—that success was assured if he used the Empire troops as his spearhead.

During the first phase of the attack on Miteiriyeh Ridge, defended by the élite of the German Army, a South Australian, Sergeant William Kibby, also of the 2/48th Battalion, won a VC.

The citation for the award, made posthumously, read:

During the attack on October 23rd, 1942, the Commander of No. 17 Platoon, to which Sergeant Kibby belonged, was killed. No sooner had Sergeant Kibby assumed command than his platoon was ordered to attack a strong enemy position holding up the advance of his company.

Sergeant Kibby immediately realised the necessity for quick decisive action, and without thought for his personal safety he dashed forward towards the enemy post firing his Tommy-gun. This rapid and courageous individual action resulted in the complete silencing of the enemy fire by the killing of three of the enemy, and the capture of twelve others. With these posts silenced, his company was then able to continue the advance.

After the capture of Trig 29 on October 26, intense enemy artillery concentrations were directed on the battalion areas which were invariably followed with counter-attacks by tanks and infantry. Throughout the attacks that culminated in the capture of Trig 29 and the reorganisation period which followed, Sergeant Kibby moved from section to section, personally directing their fire and cheering the men, despite the fact that the platoon, throughout, was suffering heavy casualties.

Several times, when under intense machine-gun fire, he went out and mended the platoon's lines of communication, thus allowing mortar concentration to be directed effectively against the attack on his company's front. His whole demeanour during this difficult phase in the operations was an inspiration to his platoon.

On the night of October 30–31st, when the battalion attacked 'ring contour' 25 behind the enemy lines, it was necessary for No. 17 Platoon to move through most withering enemy machine-gun fire in order to reach its objective. These conditions did not deter Sergeant Kibby from pressing forward right to the objective despite his platoon being mowed down by machine-gun fire from point-blank range.

Only pockets of resistance still remained and Sergeant Kibby went forward alone, throwing grenades to destroy the enemy now only a few yards distant. Just as success appeared certain, he was killed by a burst of machine-gun fire. Such outstanding courageous tenacity of purpose and devotion to duty was entirely responsible for the successful capture of the company's objective. His work was an inspiration to all, and he left behind him an example and memory of a soldier who fearlessly and unselfishly fought to the end to carry out his duty.

Yet another member of the same unit won a VC during Alamein. He was Private Percival Gratwick, whose citation read:

During the attack on Trig 29 at Miteiriyeh Ridge on the night of October 25-26, 1942, the company to which Private Gratwick belonged met with severe opposition from strong enemy positions which delayed capture of the company's objective and caused a considerable number of casualties. Private Gratwick's platoon was directed at these strong positions, but its advance was stopped by intense enemy fire at short range. Withering fire of all kinds killed the platoon commander, the platoon's sergeant and many other ranks and reduced the total strength of the platoon to seven.

Private Gratwick grasped the seriousness of the situation and acting on his own initiative, with utter disregard for his own safety at a time when the remainder of the platoon were pinned down, charged the nearest post and completely destroyed the enemy with hand grenades, killing among others a complete mortar crew. As soon as this task was completed and again under heavy machine-gun fire, he charged the second post with rifle and bayonet. It was from this post that the heaviest fire had been directed.

He inflicted further casualties, and was within striking distance of his objective, when he was killed by a burst of machine-gun fire. By his brave and determined action, which completely unnerved the enemy, and by his successful reduction of the enemy's strength, Private Gratwick's company was able to move forward and mop up its objective. Private Gratwick's unselfish courage, his gallant and determined efforts against the heaviest of opposition, changed a doubtful situation into the successful capture of his company's final objective.

Such actions were fairly commonplace among the Diggers at Alamein and they show why the attack succeeded. It was made by men who were entirely unable to grasp the concept of failure. They were being used as they liked being used—as a spearhead. They had artillery and armour in support, they knew the action had been carefully and intelligently planned and so they cheerfully accepted the risks and dangers.

Alamein 'made' Montgomery with the Diggers. Even though they lost heavily, they knew they were not being ruthlessly sacrificed—as had so often happened during the Great War.

Ten days after Alamein started—ten days in which the Diggers had neither washed nor shaved and hardly slept—with the German front pierced and the breakthrough in progress, Montgomery sent a message to General Morshead:

> I want to congratulate you on the magnificent work your division has done on the right of the line. Your men are absolutely splendid and the part they have played in the battle is beyond all praise. Please tell the division that I am delighted with the way it has fought.

Lieutenant-General Sir Brian Horrocks, writing of Alamein after the war, said: 'After the battle I went to see General Morshead, the Australian commander, to congratulate him on the magnificent fighting carried out by his division. His reply was the classic understatement of all time. He said: "Thank you, General. The boys were interested".'

As always, the medical orderly and stretcher-bearer was prominent at Alamein. A notable case was that of Sergeant V. Rae, 2/17th Battalion, who won the MM for his work during the fight. Rae worked ceaselessly to find and patch up wounded and when he ran out of Australians he turned to wounded enemies.

One night Rae went out to a dugout 550 metres from his front, where he had been told a German was lying. He found a young German, but had to tear off the roof of the dugout and drag the man out, head first. He cut his clothing away, dressed his wounds and set a fractured leg and got him to an

These Diggers had packed and were about to leave Tobruk after being relieved. They gave the relieving troops a tune on their — scrounged — gramophone. (*Imperial War Museum*)

In Tobruk men of the 2/17th Battalion with a captured Italian 75-mm gun. With this and other captured guns they became known as the 'Bush Artillery'. (*Imperial War Museum*)

Curley (*Imperial War Museum*)

Kanga (*Australian War Memorial*)

ambulance. The German insisted on giving Rae his belt buckle, a wallet and money. Rae also found a wounded German officer among bodies in a trench and patched him up. 'He even shook hands,' Rae wrote later. '*That* would have made Adolph smile.'

Finally he found a seriously wounded Italian, but was machine-gunned and shelled several times on the way back with him. 'He was a bad risk,' Rae wrote. 'He'd probably die and he nearly cost me my life and the lives of others who helped me. We got him in, anyway. Poor chap, he wasn't in a condition to be grateful.'

The Diggers left the Middle East after Alamein. The 6th and 7th Divisions were already in action in the Pacific and the 9th was to follow. The 9th had helped to give impetus to the Axis retreat and had now to fight battles nearer home.

The work of the Diggers in the Middle East was summed up by Lord Alexander when he farewelled the 9th Division at a parade in Palestine. Some of his staff said later that for the first time in his career Alexander showed intense emotion at this farewelling, for he had come to have real affection for the Diggers. He said:

The Battle of Alamein has made history, and you are in the proud position of having taken a major part in that great victory. Your reputation as fighters has always been famous but I do not believe you have ever fought with greater bravery or distinction than you did during that battle when you broke the German and Italian armies in the Western Desert. Now you have added fresh lustre to your already illustrious name. . . .

. . . There is no doubt that the fortunes of war have turned in our favour. We now have the initiative and can strike when and where we will. It is we who will choose the future battlegrounds, and we will choose them where we can hit the enemy hardest—and hurt him most.

There is a hard and bitter struggle ahead before we come to final victory, and much hard fighting to be done. . . . But wherever you may be my thoughts will always go with you, and I shall follow your fortunes with interest and your successes with admiration.

There is one thought I shall cherish above all others—
under my command fought the Ninth Australian Division.

And many commanders, perhaps less articulate than Alexander, felt the same way about other Australian divisions.

The Diggers left behind them mates buried in several Mediterranean countries and right along the Egyptian and Libyan
coasts from Alexandria to Benghazi. More than 1000 of the
2284 graves at Tobruk War Cemetery are those of Australians, as are a large proportion of the 7000 at Alamein.

But memories of the AIF didn't die, nor are they likely
to.

In Lebanon details of AIF feats of arms are etched into
roadside rock-faces near Dog River. They record victories in
1918 and in 1941–42. And the pillars of the railway bridge
across Dog River (built by Australian engineers) are adorned
with the rising sun badge, a strange sight 16,000 kilometres
from home.

19

'If They Had Any Fear . . . '

During the Second World War the Digger fought in an extraordinary variety of terrain—desert, plains, mountains, beaches, jungle. He fought in bitter cold, driving rain, stinging sandstorms, prostrating heat. He experienced winter and summer campaigns and knew what it was like to be attacked by tanks, artillery and bombers. He showed that he could adapt himself to any kind of warfare, but of all types jungle warfare was the hardest, the most exhausting, the most nerve-racking.

Long before the rest of the AIF knew anything about jungle fighting the 8th Division in Malaya were veterans. But they never did get a chance to show what they could really do, despite a number of fierce actions before the surrender. Churchill in *The Second World War* noted that 'an Australian Division, after fighting extremely well in Johore, had been destroyed or captured in circumstances . . . for which British war direction was responsible'.

This was a frank admission of the ineptness of British leadership. The capture of the 8th Division was no blot on the Digger reputation; the battalions of the division fought hard until the surrender, and the spirit of the men never weakened during all the years they were in captivity.

As a fighting man the Digger of the 8th Division was just as good as the man of the other divisions. Take the case of Private Roy Brown who was lying on top of a cutting near a road the day his battalion made contact with the Japanese in Malaya.

A bomb-burst undermined the bank and Brown fell to the road, losing his rifle. Four Japanese rushed at him, stabbing at him with their daggers. Brown lashed out with his fists

and though he took fourteen stab wounds he yet managed
to wrest a dagger from one of them and attack the others.
He killed two and wounded two. A mate carried Brown part
of the way to safety, but then for five kilometres the pair
had to fight their way back. In hospital it was found that
Brown had a slight fracture of the skull as well as the stab
wounds, some of which were dangerous.

As early in the war as this Brown and others showed that
the Japanese was no more a physical superman than the Nazi
stormtrooper.

Churchill believed that the Digger reputation would keep
the Japanese away from Australia. He wrote: 'I did not believe
that Japan . . . would be likely to send an army of 150,000—
less would have been futile—to a major struggle with the
Australian nation, whose men had proved their fighting quality on every occasion on which they had been engaged.'

A fine compliment to the Digger, but a mistaken belief,
because Japan did send an army. It didn't reach Australia, but
that was due to the Digger and to the Australian and American navies.

Before New Guinea, the AIF had never fought a campaign,
as such, in the jungle. This challenging terrain tried and
tested the Digger's stamina and endurance, his mental balance
and self-discipline, his ability to fight on a one-man front.

The great bogy which had been conjured up in the popular
imagination before New Guinea was that of 'infiltration'. It
was said the Japanese could not be beaten because he would
infiltrate behind the Australian lines as he had done in Malaya; that he was dangerous because he was an experienced
jungle fighter. In 1942 many civilian Australians despaired
of the AIF holding the enemy away from the mainland.

The Digger himself was confident enough. He had beaten
Germans, Italians and Vichy French on ground they knew
better than he did. Why should the Japanese be more difficult? He was difficult because of the jungle, but the Digger
soon realised that the jungle was tough for both sides in a
campaign of movement. He also realised that if the Japanese
infiltrated behind him, then logically, he was also behind the
Japanese. Who stabbed whom in the back was therefore a

matter of relative courage and skill.

The Digger had one decided disadvantage: he was not a fanatic, while the Japanese was just that. To him death meant glory in the sight of his ancestors; the Digger didn't want to die at all. When death means nothing to a man he can be dangerous—and the Japanese soldiers were dangerous.

The Digger was fighting two wars at once in the jungle—against the Japanese *and* the country. In the end he mastered both—in Papua and New Guinea, the Solomons, Borneo, Java and elsewhere in the islands to Australia's north. One reason for his spectacular success was that for the first time in history he was fighting directly to protect his own soil and home, not, as always before, fighting for a vague ideal on somebody else's behalf; his victory in the jungle was even more inevitable than at Alamein. Another reason for success was that control of operations was now the direct concern of Australian generals.

Until New Guinea all Australians who had fought abroad were volunteers—very special men. But the Diggers of the AIF, going to New Guinea, found that militia battalions of the AMF (Australian Military Forces) were there before them, some of them fighting as doggedly as any AIF battalion. In a way, this was something of a blow to Digger pride and tradition, for the Digger felt that the militia—who had not originally volunteered to fight abroad—were muscling in on a fight that was, by right and tradition, his. It took the jungle campaigns to show that AIF and AMF were both forces of the Australians and that there was very little difference in the make-up of the men who comprised them.

The militia could and did fight; for more than a month the 39th Battalion, aided only by a few native troops, held the Japanese without AIF help. Later the 3rd Militia Battalion helped the AIF's 25th Brigade to push the Japanese back across the jungle spine of the island to the northern coasts. Still later, militia battalions fought side by side with AIF units in the campaign of the northern beaches.

'Are these men Diggers?' a Bardia veteran asked me. 'They didn't volunteer to fight anywhere abroad; so we can't accept them as Diggers.'

By the 1940s the word Digger had a broad application—
a much wider one than in 1917. Any Australian soldier who
had actively fought an enemy was a Digger. Men of the
militia infantry battalions were probably more worthy of the
title than AIF men who fought the Libyan campaigns from
base units in Cairo. Were militiamen who died in action to
be deprived of the Digger tradition? The 39th Battalion's
part in the Owen Stanleys campaign was onerous, important
and unenviable; these men were certainly Diggers.

The first AIF units in action in New Guinea were the
2/14th and 2/16th Battalions of the 21st Brigade, successful
in the Syrian campaign of the previous year. But although
fit, confident and game, they could not hold the Japanese.
Struggling to find their feet in a green hell they had never
seen before, they were pushed slowly back towards Port
Moresby. It was a tragic retreat. The sick and wounded
mingled with troops still more or less fighting fit. Only the
seriously wounded were carried; the rest were led, or they
limped, crawled or hobbled. The tracks were deep in mud,
down one hill, up the next; rain drenched the toiling men.
A man carried his mate when he himself should have been
on a stretcher. And always there was someone, one man, up
there at front—the one-man front—trying to check the Japanese advance.

The retreat went on to Ioribaiwa Ridge, where on 28
September 1942 the 7th Division's 25th Brigade (2/25th,
2/31st and 2/33rd Battalions), the 3rd Militia Battalion and
the 2/1st Pioneers held the Japanese and began to push them
back. The 6th Division's 16th Brigade (2/1st, 2/2nd and
2/3rd Battalions) reinforced the 25th Brigade and gave impetus to the enemy retreat. These battalions had fought at
Bardia, at First Tobruk, Derna, Greece and, in the case of
the 2/3rd, Syria. The Owen Stanleys campaign—the closest
to Australia ever fought—was over. But it was only the
beginning of three years of jungle warfare, for the Japanese
were difficult to dislodge.

In these campaigns, for the first time in the Second World
War, the Digger met the Doughboy and was not particularly
impressed. He admitted that the US Marines were good, but

thought little of the average infantryman, who seemed to lack dash. The Americans had all the equipment and fire-power imaginable but seldom followed it up. The Australians, with considerably less equipment, sometimes finished off American actions.

Conversely, the Americans were deeply impressed by the Australians. General R. L. Eichelberger, who commanded the US forces in New Guinea, noted in the *Saturday Evening Post* that 'the Aussies are unique soldiers, amazingly casual but willing to tackle anything. They liked to work in small parties and were always successful. If they had any fear they never showed it'.

Being human, the Digger does know fear, but it is a basic Australian characteristic not to show emotion. In any case, the fear is only in the anticipation; once an action has started the Digger has neither time nor inclination for fear. An American war correspondent, reporting from Milne Bay, said he was prepared to admit that the Australian soldier 'must be equal to the world's best and is probably *the* best'.

> These men [he wrote] have a high degree of inner self-discipline. When they set out on an action there is never any fuss; the officer or sergeant gives a few quiet orders, the men make a gag about something and off they go. There is a hard brightness about their eyes and they look as if they mean business. They do the job and come back— perhaps minus a man. They don't talk about him, but you can see they're thinking about him. They smoke a cigarette with steady hands, talk a little about the action and then say: 'Well, what sort of stunt is next on the list?'

Once he had adapted himself to jungle warfare, the Digger found that it suited him. It was made to order for section and platoon actions, for aggressive patrolling and individual initiative. The Australians who had gone to the Boer War would have liked this new warfare. The terms 'division' and 'brigade' had almost no meaning and battalion actions were rare.

One of the outstanding jungle actions was the 2/17th Battalion's fight at Jivevaneng. Brigadier Windeyer (later

Major-General, CB, CBE, DSO and bar) described it as the finest achievement of any Australian battalion in the Pacific war. Opinions might vary on this, but the 2/17th's battle was noteworthy enough, if only as an example of typical jungle warfare.

The Jivevaneng fight began, for practical purposes, on 11 October 1943. A and B Companies under Major Maclarn were in brigade reserve and C and D, led by Lieutenant-Colonel Simpson, moved inland towards the village of Kumawa, but ran into a strong force. A and C Companies, brought forward to Jivevaneng, also met resistance. Eventually the two forces joined at Jivevaneng for the drive up the main Sattelberg track.

One of the most noteworthy of several actions here was an attack on the Knoll by two platoons of B Company led by Captain T. C. Sheldon. No. 10 Platoon had to charge straight up a steep slope amid a shower of Japanese grenades. Grenades thrown by the Diggers rolled back down the hill exploding among them, but they overran the hill and held it against repeated counter-attacks. Lieutenant R. J. Bennie, MC, had led his platoon to the objective, throwing hand grenades and shooting two Japanese. When Captain Sheldon was wounded Lieutenant Bennie took command.

Private R. C. Brooks, MM, showed great coolness and courage in killing all the Japanese occupying three well-defended posts; his courage and dash demoralised the enemy on that part of the Knoll. Acting-Sergeant J. N. Wood, MM, whose platoon held a vital part of the perimeter, dashed forward under heavy machine-gun fire and grenades to man a Bren-gun when the gunner was wounded. The battalion lost three killed and ten wounded in the assault on the Knoll.

The real fight for Jivevaneng began on 16 October when it became apparent that the Japanese were launching an all-out offensive to recapture Scarlet Beach and Finschhafen. A fierce attack on Battalion HQ was followed by heavy shelling of the mortar and machine-gun platoon areas. Four more attacks were repulsed. Private R. A. Gillespie fought it out with a Japanese machine-gun crew at forty-five metres and destroyed the gun and crew. The RAP came under direct fire and Private McCarthy, the padre's batman, was wounded and his leg had to be amputated.

Shells from a small Japanese mountain-gun, firing from fifty-five metres away, were bursting above the Australians. Sergeant T. Johnson was fatally wounded when shot by a patrol. He was one of four men killed and fourteen wounded on the first day.

Lance-Corporal M. O'Brien, seriously wounded while leading a small patrol, had not been found that night, despite an intensive search. Two other men on patrol with him were killed. Next night the Japanese cut the track and the battalion was virtually besieged, with the Japanese dug in only about fifty metres from HQ.

By 21 October supplies were coming in over a rough track through New Tareko, but 120 native carriers, though working hard, could not bring up all the supplies needed. That afternoon twenty-four Liberators escorted by sixteen Lightnings bombed and strafed the Sattelberg–Wareo area. Their 214 tonnes of bombs was probably the biggest bombing of the Pacific war to that time. The battalion's best meal that day consisted of three biscuits each, covered with jam.

Within the next few days small parties of left-out-of-battle men were coming up each day through New Tareko and other men were going back for a rest. On 24 October there was a clash; and Corporal W. K. Bartlett was killed. Next day Private D. Garrick fought a machine-gun duel at about ten metres range. Many forward posts were only about twenty-five metres from the Japanese and casualties mounted daily. Thickness of jungle, heavy rain and steep ground impeded the battalion's patrols.

Earlier training in handling casualties was proving its worth, for all seriously wounded cases had to be kept for varying periods in the lines, and the stretcher-bearers attached to each company were able to keep the sulpha and atebrin going, and also to control the dysentery with sulphaguanadine. Dysentery was a real threat because of the hordes of blowflies which infested the area.

On 30 October a Japanese patrol walked into an ambush set at Cemetery Corner, losing four killed and three wounded.

The Japanese were rather more patient than the Diggers. Again and again a lone Japanese would edge forward and lie silent and still for hours, waiting for a chance to shoot anybody who might show himself for a moment.

Next day a Japanese division made a major attack on Scarlet Beach. The fresh 24th and 26th Brigades killed hundreds of Japanese in decisively repulsing the attacks. The 2/17th's casualties during October were twenty-four killed, fifty wounded and forty-nine evacuated sick, mostly with malaria.

On 30 and 31 October the battalion made two attempts to open the track as a preliminary to pressing on to the 900-metre peak of Sattelberg. They had eighteen casualties. On 3 November they linked with the 2/13th Battalion at Cemetery Corner and opened the Sattelberg track. C Company (Captain Dinning) was ordered to dislodge a Japanese force threatening the track leading past Jivevaneng. The company tried to keep going against heavy fire, but suffered heavy casualties, and went to ground eighteen metres from the enemy positions, where it became engaged in one of the fiercest minor battles of the war.

Private D. McCaffery saw a head appear at a peep-hole six metres away. He fired at it and was certain he scored a hit. A head reappeared and he fired again, only to have a third head pop up. McCaffery was disgusted with his poor marksmanship, but three dead Japanese were later found in the tunnel leading to the peep-hole.

Vickers and Brens poured thousands of rounds into the enemy positions and Bren barrels became so hot they had to be replaced. An hour and a half after the fight started twenty-five Japanese armed with a 'woodpecker' machine-gun counter-attacked a forward section but all were killed by Bren- and Owen-gunners.

The conduct of Lance-Sergeant B. G. Dawes, DCM, was outstanding. Soon after the attack started one of his section was killed and four others, including the platoon commander, Lieutenant Graham, wounded. Although only about fifteen metres from the enemy, Lance-Sergeant Dawes dragged all casualties to safety and arranged their evacuation while maintaining the rest of his section in action. Soon after his section was heavily attacked. He waited until the Japanese were ten metres away before giving the order to fire. The attack broke up. Then Lance-Sergeant Dawes established himself in a forward position only seven metres from the Japanese and despite flooding, rain and much sniping he harassed the enemy for thirty-six hours. He threw sixty grenades and his aggressive example helped force the enemy to withdraw.

Corporal I. Moore, MM, of 15 Platoon, on his own initiative, walked through heavy fire several times to carry Bren barrels forward and to help wounded back to battalion headquarters. He carried Lieutenant Graham back on his shoulder, but the lieutenant died of his wounds. Sergeant W. Pearce, MM, of 15 Platoon, also did gallant work in supplying the forward platoons with ammunition.

The Japanese fired forty bombs from an 81-millimetre mortar, killing two men and wounding three, making the company's casualties for the day eight killed and eleven wounded. Tropical rain flooded their positions by night and the men huddled in muddy slit trenches, with the Japanese in position as close as three metres away.

At sunrise on the second day Corporal F. G. Mahalm was fatally wounded by a sniper while carrying food to his section. The fight continued all day without gain to either side, but next morning patrols found that the Japanese had withdrawn. The battalion was able to hand over the important sector to the 2/24th Battalion clear of all enemy.

The battalion's strength when it landed at Scarlet Beach was thirty-one officers and 703 men; it left Jivevaneng with twenty officers and 437 men, including many reinforcements. Most of the men thought that Jivevaneng was much worse than Tobruk. Evacuations from malaria and dengue fever reached 209 for the month.

The little war within a war was similar to many actions fought by nearly all the infantry battalions. It was a type of warfare completely different from the deserts of Libya and the mountains of Syria, with the enemy, unseen, only a few metres away. Such fighting called for particular endurance and courage and the Diggers were not found wanting.

In November 1943, among the jungle-matted hills around

Sattelberg, one of the most outstanding Australian soldiers of any war won a VC for a remarkably aggressive and considered exploit. He was Sergeant Tom Derrick, yet another member of the decoration-studded 2/48th Battalion.

Derrick already had the DCM, awarded for leadership and personal courage in action during the initial fighting at Tel el Eisa, Libya, in July 1942, and for devotion to duty for the period May to October 1942. His DCM citation read:

> Sergeant Derrick has frequently shown outstanding leadership in action. During the attack on Tel el Eisa in the early morning of 10 July 1942, Derrick, by his own personal courage and leadership, attacked and captured three Fiat machine-guns. He was personally responsible for the capture of 100 enemy by his cool determination, leading his men with great dash and bravery.
>
> Later that same night, in a counter-attack on enemy tanks and infantry at Tel el Eisa railway station, Sergeant Derrick was again outstanding in fighting qualities. He attacked and damaged two German tanks with sticky bombs and was a great factor in the successful action which restored the station to our forces. On all occasions, both in and out of action, Sergeant Derrick has been exemplary in his conduct and courage. He has proved himself to be a fine leader and brave soldier, always inspiring his men to follow his example.

Derrick was offered a commission in Libya, but he preferred to remain a sergeant. The lesser rank in no way hobbled his ability to lead. On 23 November 1943, Derrick assumed command of No. 11 Platoon of his battalion, which was trying to take the Japanese stronghold of Sattelberg. Because of the harsh terrain the only possible approach to the town lay through an open kunai (long grass) patch directly beneath the top of a cliff held by the enemy. For two hours the Australians made many attempts to climb the slopes to their objective, only to be stopped by intense machine-gun fire and grenades.

Shortly before last light Derrick's company commander reported to his CO that he had no hope of taking the vital ground and probably he would be forced off the ground he had already occupied. The company was ordered to withdraw and had actually commenced to do so when Sergeant Derrick asked permission to make a last attempt to reach the objective. His request was granted.

Moving ahead of his forward section, Derrick personally destroyed with grenades an enemy post which had been holding up his section. He then ordered his second section around on the flank. This section came under heavy fire from light machine-guns and grenades from six enemy posts. Derrick clambered well ahead of the leading men of his section and threw grenade after grenade. The Japanese bolted, leaving weapons and grenades.

By this action alone, Derrick's company gained its first foothold on the precipitous ground, but Derrick wasn't finished yet. With his first and third sections he went on to deal with the remaining three enemy posts in the area. Four times he ran forward and threw grenades at a range of 5½ to seven metres and silenced these posts. In all, Derrick reduced ten enemy positions. From the vital ground that he had captured the remainder of the battalion moved on to capture Sattelberg the following morning.

Derrick's VC citation read, in part:

> Undoubtedly his outstanding gallantry, fine leadership and refusal to admit defeat resulted in the capture of Sattelberg. His gallantry and thoroughness were an inspiration to his platoon and to his company and have served as a conspicuous example of fearless devotion to duty throughout the whole battalion.

What the citation failed to emphasise was that Derrick's action was not the result of a fight-out from some personally desperate situation, nor was it an impulsive heat-of-the-moment exploit. He *asked* permission to make the attack—a calculated and considered one. He took the main risks himself and did not needlessly expose his men. Because of this his VC was all the more worthy. I regard his exploit at Sattelberg as the most brilliant individual action by an Australian infantryman during World War II.

Returning to Australia, Derrick now did accept a commis-

sion. The army likes to keep a VC winner safe and Derrick
could have had a job in Australia. But Derrick objected; he
had enlisted to fight, he said. He stayed with his battalion.

By this time the Australians had cleared the Japanese from
most of the islands and the war ended as they were driving
them from Borneo. Tom Derrick didn't see the end of the
war—he was killed in action in Borneo. And as always, he
was leading his men—from the front.

Because of men such as Derrick it was not surprising that
when the second AIF ceased to exist the Digger had proved
himself as good as the Diggers of the Great War and had
shown that he possessed guts, endurance, resilience, audacity,
enterprise, adaptability, resourcefulness and dash.

Many Diggers had died to build up this reputation—but
deaths were not nearly as many as in the Great War. The
AIF and certain militia units which fought in New Guinea
had a total of 21,558 deaths—killed in action, died of wounds
or injuries. (The RAAF had 10,264 deaths and the RAN
2004.) Another 177,049 had been wounded. Some men were
wounded several times. Prisoners escaped or repatriated
totalled 20,920; most of these men were from the 8th Division. Another 222 Diggers were lost when a Japanese
submarine sank the hospital ship *Centaur* off the east coast of
Australia.

20

'Such Gallantry'

After the Diggers returned home from World War II there was no longer an AIF by that name and it was doubtful if the name would ever again be used. The word 'imperial' had become unpopular because of its connotations of colonialism and empire. At no time had Diggers worn a badge with the words 'Australian Imperial Force'; the rising sun badge between 1903 and 1949 carried the scroll legend 'AUSTRALIAN COMMONWEALTH MILITARY FORCES'.

The AIF legend continued, however, and the post-1945 soldier was not so very different from his predecessors. Some veterans stayed in the army though the general public held the patronising view that the only men who joined the army in peace time were those who could not make a living in any other way.

Australia was at peace but at the end of the war it was considered politically desirable to have an Australian force as part of the Allied army of occupation in Japan. In 1948 a brigade group of three battalions of regular soldiers was formed; the infantry units were designated 1st, 2nd and 3rd Battalions of the Australian Regiment. A year later this formation became the Royal Australian Regiment (RAR) and the word 'Commonwealth' was dropped from the rising sun badge; now it was 'AUSTRALIAN MILITARY FORCES'.

In 1950 a United Nations army was hurriedly assembled to defend South Korea when the Chinese and North Koreans attacked it. The 3rd Battalion RAR under Lieut.-Colonel C. H. Green was sent to Korea as part of a British Commonwealth Division.

'The Australians were hungry for a fight,' wrote British correspondent Reginald Thompson. 'They had volunteered

for adventure and so far they had not had it. They looked
bronzed, fine soldiers, keen-eyed under the slouch hats for
which GIs were offering twenty dollars and no takers.'

These new Diggers were involved in some mopping-up
operations and were then made part of the force which was
to invade North Korea. About thirteen kilometres north of
Sariwon the 3rd Battalion came up against strong enemy
positions where the Diggers achieved an impressive victory—
without firing a shot. Lieut.-Colonel Green could see that a
conventional attack would certainly result in the loss of many
Australian lives and it might also be abortive. Two company
commanders, Majors G. M. Thirlwell and I. B. Ferguson,
mounted a tank and with an interpreter drove directly up to
the enemy position. Here they explained to the North Koreans that they were surrounded and heavily outnumbered
and so should give up and save their lives. In this way 1982
North Koreans surrendered, with large quantities of arms, to
an Australian unit less than half their strength. The 3rd
Battalion was later described by a senior British officer as
'the finest fighting infantry battalion I have ever seen'. At
Sariwon its officers demonstrated finesse as well.

Under Green's impressive leadership, the 3rd pressed on
and on 22 October 1950 made a bayonet attack against
entrenched North Koreans; at a cost to themselves of seven
wounded they killed 270 and captured 239 enemy. Brigadier
B. A. Coad, British commander of the Australian units in
Korea at that time, witnessed what he termed 'a marvellous
sight'. As he reported, 'An Australian platoon lined up along
a paddy field and walked through it as though they were
driving snipe rather than enemy soldiers. The Australians
thoroughly enjoyed it, they did the whole day and they were
absolutely in their element.'

Reginald Thompson saw the Australians at this time when
they were trying to 'bash their way forward'. He wrote,
'These men impressed us with their magnificent health. Smart,
soldierly in bearing, shaven, bronzed, they provided a remarkable contrast with the drab-booted, stubble-chinned
American troops, slouching by the roadsides when forced to
their feet and looking like convicts in their drab denims.'

After further hard fighting, Lieut.-Colonel Green was killed by a shell fragment which pierced his tent and hit him while he slept, and Lieut.-Colonel I. B. Ferguson became the CO. Soon, however, the UN troops were withdrawing under an overwhelming winter enemy counter-offensive and the Diggers, with the rest of the Commonwealth Brigade, dug in on 'Frostbite Ridge'.

Here they fought what some of them thought was a Western Front type of war, with clashes in no-man's-land. In fact, it was neither as severe nor as unpleasant nor as protracted as conditions on the Western Front had been, but Thompson, who saw the Australians during the hard winter of 1950–51, thought that they suffered more than any other troops.

> In a way they were the toughest and most experienced soldiers [he wrote], but many of them were over thirty-five years of age and some, having cheated on their age, were over forty. The cold was beginning to find them out and the doctors were working hard, examining men doubled up with lumbago and kidney trouble. But the men were remarkably cheerful and making themselves snug in rice straw.

On 22 April 1951 the Chinese struck an avalanche-like blow at the UN positions astride the Kapyong River, where the Diggers occupied Hill 504. The South Korean line fell to pieces under the onslaught and many South Korean soldiers, fleeing for their lives, passed through the Australian lines. Some North Koreans posing as South Koreans infiltrated at the same time, thus getting behind the Australians. Then the Chinese frontally attacked in waves. The forward companies were soon outflanked by the Chinese, who drenched the position with mortar and machine-gun fire. The Diggers retaliated with Brens, mortars and rifles, while behind them New Zealand artillerymen fired their 25-pounder shells as fast as they could load them. When the sun went down on Anzac Day (25 April) thirty Australians lay dead or mortally wounded but they had killed 600 North Koreans and Chinese and held their ground.

Some of the troops commented wrily that the fight had been specially arranged for them as an Anzac Day celebration.

The 3rd Battalion received a US Presidential Citation for the action at Kapyong. The Citation read:

> The seriousness of the enemy breakthrough had been changed from defeat to victory by the gallant stand of these courageous soldiers. . . . They displayed such gallantry, determination and *esprit de corps* in accomplishing their mission as to set them apart from and above other units in the campaign, and by their achievements they have brought distinguished credit to themselves and their homeland.

On the Imjin River front, in October 1951, the 3rd Battalion was heavily involved in attacks on Hills 317 and 217. That on Hill 317, a key point in the Chinese winter line, was as fierce as anything in Australian experience. The hill was pyramid-shaped and so steep that it could be climbed only on hands and knees. From their apparently impregnable position the Chinese tried to smother the attack with heavy machine-gun fire. But the Diggers, with supporting artillery fire and air-strikes just ahead of their advance, inched their way forward. As dusk fell they clawed their way onto the plateau and in a bayonet charge routed the Chinese who broke and fled, leaving sixty-eight dead. On Hill 217 the Battalion withstood repeated counter-attacks throughout the night of 8–9 October and again the Chinese fled, leaving 120 dead.

The 3rd Battalion's own commander, Lieut.-Colonel F. G. Hassett (who had succeeded Ferguson), said, 'Their sheer guts is beyond belief.' It is most unusual for a CO to pay his own unit such a soldierly compliment—because it is better left to outsiders—so Hassett's comment is all the more impressive.

In thirty-seven months all three regular battalions saw service in Korea and enhanced an already impressive reputation. By the time the armistice was signed in 1953, 278 Diggers had been killed in action.

21

No Farewell, No Welcome

I t was becoming evident that Australians would be doing their soldiering, for the foreseeable future, in south-east Asia and the Pacific. Only two years after the withdrawal of the Royal Australian Regiment from Korea it was off to war again. This time the theatre of operations was Malaya, where the 'Emergency' had begun in 1948. As in Korea, all the men who went there were volunteer regulars. And as with Korea—also with Vietnam, still to come—there was no ecstatic public farewell. Many Australians did not know exactly what the 'Malayan Emergency' was. Others said, 'It's a British affair, let Australia stay out of it.'

In fact, the Emergency occurred because communists tried to take over the Malay peninsula and prevent it from becoming an independent, democratic republic. The Australians' task for five years was the unpublicised but exhausting one of patrolling the 'pacified' area. When the Emergency ended in 1960 fifteen Australians had been killed and twenty-seven wounded.

The idea of establishing a Federation of Malaysia provoked Indonesia into what it called 'Konfrontasi', a word taken into Australian military history as 'Confrontation'. In April 1963 Australian troops from the Far East Strategic Reserve helped defend Borneo against Indonesian guerrillas raiding Sabah and Sarawak. The 3rd and 4th Battalion RAR, with their own artillery and engineering support, were responsible for a front of fifty-six kilometres to a depth of forty-eight kilometres. Operating in such a jungle area with a debilitating climate was an exacting type of warfare. The whole campaign was one of jungle patrols and scouting and it called for physical and mental toughness of a high order. It was a form of peaceful penetration and similar in some ways to the aggressive patrolling by the 9th Division out from Tobruk in 1941.

The 'Emergency' and the 'Confrontation' were rather like preludes to what might well be called the 'Crisis'—the war in Vietnam. The emotions aroused by Australian involvement in this war were profound, violent and no doubt sometimes sincere. Most of those who rejected the political decision to send Australian soldiers seemed also to reject the men themselves. This was unfair and illogical. To vilify and condemn men who have already been subjected to the fear and horror of war seems peculiarly perverse. To blame the Americans for all the evils which befell Vietnam is also peculiarly ignorant; the North Vietnamese were guilty of atrocities and massacres long before the US became involved in the war.

The Vietnam War proved, as no other modern conflict has done, that war brutalises. The brutality of war, for the first time, came into people's living-rooms—via the television set. For the Australian public to abuse Australian soldiers for fighting in this war is a strange and spurious form of holier-than-thou piety. Rather, the soldiers deserve sympathy and compassion for their ordeal.

Whatever the critics say, the men concerned—national servicemen as well as regulars—were as a matter of historic fact courageous, competent and professional. No war correspondent or student of war has suggested that Australian troops were involved in any atrocity or attack on civilians.

The first Australians to serve in Vietnam were thirty members of the Australian Army Training Team (AATT) posted there in 1962. Three years later three battalions with supporting units were on service, south-east of Saigon, a difficult jungle area. The AATT built for itself a unique place in Australian military history; for its size it is the most decorated unit. In addition to four Victoria Crosses, it was awarded the American Meritorious Unit Citation, the Vietnamese Gallantry Citation and many personal awards for bravery.

The first of these four VCs was awarded for an example of courage typical in the Australian Army since the Boer War and for an act of a type repeated over and over, during World

War I particularly. Warrant Officers K. A. Wheatley and Swanton went out with a South Vietnamese company on 13 November 1965 on a search-and-destroy mission. The two Australians were with the right platoon, which came under Viet Cong fire in open rice fields. Wheatley radioed that Swanton had been hit in the chest and asked for an air-strike and aircraft to evacuate casualties.

His platoon of Vietnamese began to break under fire; the medical orderly told Wheatley that Swanton was dying and then he ran. Wheatley half-carried, half-dragged Swanton towards the jungle nearly 230 metres away, rejecting the wounded man's pleas that he get away to safety. As the Viet Cong came running through the jungle Wheatley pulled the pins from two grenades and calmly awaited the enemy. Soon afterwards two grenade explosions were heard and then several bursts of fire. The two Australian bodies were found next morning. Wheatley had obeyed the creed of mateship.

The other three VC winners in Vietnam were: Major P. J. Badcoe (posthumous) for acts of gallantry on 23 February, 7 March and 7 April 1967; Warrant Officer K. Payne, 24 May 1969; and Warrant Officer R. S. Simpson 6 May, 11 May 1969.

The Australian approach to warfare in Vietnam was completely different from that of the Americans. The Americans usually preceded their infantry attacks with massive aerial bombardment. Whether they did this or not they then moved through the jungle in large, closely bunched groups, often noisily; they hoped to defeat any ambush by superior firepower. But as the bombs fell the Viet Cong, still undefeated, hurried out of an area. Mostly the Americans did not see or hear the enemy, who simply backed away from advancing patrols until they themselves were ready to strike. The enemy attacks were nearly always successful.

The Australians fought the Viet Cong in the way that the 2nd AIF and the militia battalions had fought the Japanese. Their small patrols spread out and moved quietly through the jungle. If one man reported contact with the Viet Cong his mates would close on him rapidly to attack the enemy from all sides. Their success proved the Australian methods to be the best. Brigadier T. F. Serong of the AATT said, 'Conventional soldiers think of the jungle as being full of lurking enemies. Under our system, *we* do the lurking.'

The Australian Special Air Service, raised in July 1957, fought in Indonesia and Vietnam. It is not regarded as an élite force—as is the British SAS—just as a specialist one. With great experience in modern warfare the Australian SAS is designed to gain maximum information from limited resources and to help in the application of 'economy of force'. An example of their successful technique occurred in Vietnam.

Over a period of six weeks six SAS patrols carried out constant surveillance over 15,000 metres of enemy approaches to the Australian base. Twice-daily reports by radio from each patrol kept HQ informed. In this way the whole Australian force could take part in a large-scale sweep without being worried about having an open flank. Only one infantry company was left behind to defend the base.

An SAS patrol in Vietnam could operate for fourteen days on its own supplies. This remarkable endurance enabled the SAS to move slowly and stealthily into enemy territory, where they found the Viet Cong noisy and careless; they were confident that their enemies could not get near to their bases. SAS patrols penetrated right into Viet Cong base areas for close reconnaissance. They sprang ambushes with devastating success and killed at least 500 enemy for the loss of one Australian. It was an impressive military performance and when General William Westmoreland commanded Allied forces in Vietnam he ordered a training school to be established so that Americans and other troops could be taught ASAS methods.

The most important Australian battle was that of Long Tan in August 1966. D Company (108 men) of 6th Battalion of the Royal Australian Regiment, under Major R. Smith, advanced into a Viet Cong trap. As two platoons moved, under torrential rain, through a rubber plantation, the ambush was sprung and the Australians were hit by mortar and small-arms fire. They did not then know it but 2500 enemy opposed them.

The Australians drew into a circle, hastily dug pits for shelter and with cool deliberation stopped wave after wave of Viet Cong attacks. With their ammunition running low they were reinforced by A Company, led by the Battalion's CO, Lieut.-Colonel Townshend. After three hours of incessant combat the Viet Cong ran, leaving behind 245 dead; they probably suffered 350 wounded. The Australians lost seventeen dead and had twenty-one men wounded.

One of the most remarkable Diggers thrown up by the Vietnam campaign was Captain Barry Petersen, who was posted to the Training Team in August 1963. Petersen was given the task of supporting and supervising the operations of paramilitary teams of Montagnards in Darlac province in the southern central highlands. *Montagnards* was the collective term given by the French, when they controlled Indo-China, to the primitive mountain tribes of the peninsula.

For two years Petersen led an adventurous and dangerous life. He gained acceptance as a Montagnard chief in his own right and he was the central figure in quelling a widespread tribal revolt against the Vietnamese government. He recruited, organised and trained an army of well over one thousand men, and he was involved in the training of men for Operation Delta, a raid by Vietnamese special forces into North Vietnam. Having had thirty months experience against communist terrorists in Malaya, Petersen knew that the most effective way of stopping the Viet Cong fighters from moving along the jungle trails was to saturate all possible routes with small, well-trained patrols of men familiar with the area— and this worked.

The Vietnamese renamed Petersen's teams the Truong Son force and gave him a special badge in the form of a tiger for their berets. The Montagnards held Petersen in high regard and affection and were loyal to him. They gave him the tribal name of Dam San, after a legendary warrior. Besides paying all his people—from American money—Petersen financed any logistic support which was not provided direct from HQ in Saigon. To do this he had the enormous fund of five million Vietnamese piastres in cash. Internal Vietnamese jealousies led to Petersen's removal from the high-

lands in October 1965 by which time he had become a
legend among the Montagnards. In May 1970 he returned,
as a major, for a one-year tour of duty as a company commander with the 2nd Battalion RAR.

Lieutenant-General Sir Thomas Daly, the Australian Army
chief for most of the Vietnam War, sent a letter to the CO
of the Australian Army Training Team on 30 July 1969. 'It
is always rewarding talking to these chaps [the members of
the Team],' he wrote. 'They are the salt of the earth and all
those who know them cannot but be inspired by the tremendous job they are doing for Vietnam and Australia.'

Vietnam turned sour but the critics of the Australian effort
must realise that the Diggers in Vietnam were all 'doing a
job for Australia'.

But they were not welcomed home by cheering crowds
and one woman, having smeared herself with red paint,
attacked the CO of the 1st Battalion as he was leading a
parade in Sydney. He was wearing the Distinguished Service
Order for his courage and leadership.

Vietnam was Australia's longest war, lasting over ten years.
About 47,000 soldiers served during the war, with a strength
of 8000 at the peak of commitment. Total casualties were
415 killed and 2348 wounded.

Since this book is about Diggers on active service I will
not go into changes in army organisation since the end of
Australia's involvement in the Vietnam War in 1972. How-
ever, it is worth noting that in 1970 the 'AUSTRALIAN MILI-
TARY FORCES' disappeared from the rising sun badge and was
replaced by the one word, 'AUSTRALIA'. As a shoulder title
'AUSTRALIA' has been worn by hundreds of thousands of
Diggers and it says all that needs to be said.

It seems evident that when Australian soldiers are again
called upon to fight they will produce the same high courage
and the same standards of enterprise, élan and initiative established during the Boer War and confirmed during the
Great War. And, of course, mateship.

22

Leading by Example

Brigadier-General Sir James Edmonds, in his book *Military Operations France and Belgium, 1918*, analysed leadership in the British Empire armies:

> The leading of the Canadian and Australian officers and NCOs was superior to that of the British regimental cadres, and no doubt for the reason that they had been selected for their practical experience and power over men and not for theoretical proficiency and general education.

In 1918 the AIF was contributing contingents of officers to the British Army because the British could no longer find enough men capable of holding commissions. Even in 1940 the British official historian of the Great War was telling the British Army that the AIF of the Great War was a model corps.

It has always been rare for a British private soldier to gain commissioned rank. By contrast, the great majority of AIF officers had commenced as privates. Many of them succeeded to the command of battalions and more than a few became brigade commanders. A typical case was that of A. S. Blackburn, VC, a private at Gallipoli, a brigadier in World War II. Lieutenant-General Sir Stanley Savige, whose adventures in Mesopotamia are described in Chapter Fourteen, was an NCO at Gallipoli.

There has never been a shortage of competent officers in the AIF. It is true that some reinforcement officers, commissioned from an officers' training school and not in the field, were found to be unsuitable, but never were they permitted to retain control of fighting men. Such officers were seconded to duty in units where their lack of leadership could not endanger their men, though many a time a new reinforcement officer has been nursed through his early actions by platoon NCOs who could see that he 'had the makings'.

Senior British leaders have gone on record as saying that Australian officers have not enough control over their men. In making such a statement they did not understand the relationship between an AIF officer and his men. The officer's 'control' is complete though he might exercise it in a way that many a British general, at least of the pre-1950 period, might not comprehend. He does not shout, he seldom punishes, he is never 'superior', he never drives. Above all else, the Australian officer is a leader; he leads by example. I knew one officer, a well-loved platoon commander, who used to take his own rifle on parade. (Platoon commanders were not usually armed with rifles, but this man preferred the .303 to a pistol or submachine-gun.) When, on inspection, he found a soldier with an unclean rifle the officer would give the soldier his own rifle to inspect. 'Take a look at that,' he would say. 'Yours has to be as clean as that.'

This was typical of the unwritten creed of AIF officers that they would never ask a man to do anything they themselves could not or would not do. In any infantry platoon the lieutenant is the toughest man of all, except perhaps for his platoon sergeant, who eventually would probably get a commission himself.

The Digger officer gets results in a quiet way.

I once saw an English major try to stop an Australian two-up game. He came upon a circle of Diggers behind some tents, but though they saw him approach they took no notice. It takes a lot to distract the attention of a two-up school.

'Move away from here, you men,' the major said officiously. 'This sort of thing isn't allowed.'

The Australians went on playing as if the major wasn't there. He fumed. 'Attention!' he shouted.

The Australians went on playing but came to attention as ordered. 'A quid in the guts! ... Ten bob he tails 'em. ... Come in spinner!'

'That's enough, I say!' the major ordered. 'If you don't move away from here I'll place you under arrest.'

'Pull your head in, Jack,' somebody said.

'Who said that?' the major demanded.

The thing might have become serious had not an Australian lieutenant (who had once been a private) walked by at that moment. He summed up the situation, pushed through the circle of Diggers and said quietly: 'All right, blokes, break it up.'

The circle broke up instantly and the men went off without a murmur. The lieutenant smiled politely at the major, saluted him and went about his business. It wasn't the major's fault—he just didn't know that being an officer to Australian troops is quite different from leading other troops.

Very senior officers have set personal examples. During the dark days in Greece in April 1941 when the Australians and New Zealanders were fighting a rear-guard action while extricating themselves from the country, Lieutenant-General Sir Iven Mackay and Lieutenant-General Sir Bernard Freyberg, VC (commanding the New Zealanders) particularly, were taking pains to encourage by their example a disregard of the frequent German air attacks.

> On this and the succeeding days [the *Official History* records] each of these commanders was seen at his full stature. Their influence seeped down the force. Some saw Freyberg standing nonchalantly and alone while an aircraft machine-gunned him, and missed; or saw him coolly clearing a traffic jam among the truck of a withdrawing infantry rear-guard. Others heard of Mackay sitting in the open, apparently unconcerned, during an air attack, or of Mackay waiting at the rear-guard position . . . during an air attack that lasted two and a half-hours, when his car was hit and his driver wounded.

Not a few Australian officers have been killed trying to help their men. This was the fate of Second-Lieutenant Frederick Birks, who won a VC in France shortly before his death in 1917. Accompanied by a corporal, Birks rushed a strong-point holding up an advance. When a bomb wounded the corporal, Birks went on alone, killed the Germans with the machine-gun and captured the gun. He then organised a small party and attacked another strong-point occupied by twenty-five Germans. Birks himself killed several of the enemy and captured fifteen of them. He showed wonderful coolness and courage and he performed that best of all tasks—keeping his men in splendid spirits. Not long after this, some of Birks' men were buried by a shell. He was trying to dig them out, under fire, when a second shell killed him.

Men always know the capabilities of their officers, for they have seen them fight their way to a commission. This leads to immense confidence.

General Birdwood, who came to know the Australians very well, was the first to insist that the officers must see to their men's interests before their own. This edict caused some resentment at first, but only briefly. It became an established rule of the AIF that officers and NCOs must see that their men were fed before they themselves ate. When in action platoon commanders eat the same food as their men, with their men. Australians have been astonished many times to see British officers even leave front-line posts to dine at their regular mess, often established and maintained at great effort by a special messing staff.

Social distinction between officer and men has never existed in the AIF. An officer holds the regard of his men by sheer force of personality and ability. The Digger is not easily fooled. He can take one look at an officer and know at once what sort of leader he is. No amount of bluff and bounce will deceive him. It has been said often enough that the Digger resents discipline, yet I have seen soldiers endure the most rigorous training and discipline at the hands of an officer or an NCO they respect.

In January 1943 thousands of troops were resting in the Atherton Tablelands in northern Queensland. They had returned from operations in Papua-New Guinea and were preparing for further action in the islands. In the heat of the evening a sergeant (later an officer) marched in with his thirty men after a day's hard training in the Queensland jungle. In his opinion they had not come up to scratch and he told them so. Some of these men were originals of the

6th Division; most had fought in some campaign or other. More importantly, the sergeant had four campaigns behind him.

'You're a bunch of bloody no-hopers,' he said, among other things. 'Soldiers! You aren't soldiers' bootlaces. Girl Guides could do better than you did today. Maybe you want to get yourselves killed and that's your business, but I don't want anybody saying *I* killed you. So now we're going to do two hours' drill. I'm going to march you until I wear your legs down to stumps!'

The longer he worked the platoon the better they marched and drilled. Sweat rolled off them, streaking their mud-caked faces. In a short while hundreds of off-duty troops were watching them.

Finally, after dark, the sergeant dismissed the platoon. The men were dispersing when an onlooker said to one of the sufferers: 'Jeez, what a bastard *he* is!'

The sufferer turned, dropped his rifle and threw two heavy fists into the speaker's face. 'Want to make something of it, mate?' he said. 'Any time you want a stoush come along to 14 Platoon and call Sergeant Baker a bastard.'

The incident didn't end there. An hour later the sergeant sent the platoon a dozen bottles of cold beer from the sergeants' mess. Such things as this have endeared officers and NCOs to their men ever since Australia had an army.

The quality of Digger junior leaders was never more evident than in patrol work, which calls for a high degree of initiative and leadership and strong control of men. Of all forms of warfare the Digger has always like patrolling best of all because it gives scope for initiative, enterprise and individual daring. It has brought to light many outstanding NCOs and junior officers. Patrolling, by its very nature, demands a small force and for this reason most patrols are led by a lieutenant or sergeant.

A remarkable patrol, forgotten now, was carried out by an officer of the 14th Battalion, Lieutenant Harry Danman, on the Western Front in 1917. He was detailed to occupy a certain section of trench but when he moved his platoon in he found part of the trench occupied by Germans. The

Australians fought them out and settled in, but late in the afternoon Danman got an order saying: 'Make arrangements to send a patrol to raid the enemy trenches tonight for the purpose of securing a prisoner for interrogation.'

Danman made his arrangements and the patrol moved off. Half an hour later he reported to Battalion Headquarters with the required prisoner, handed him over and walked away. He did not think it worth mentioning that the patrol had consisted of himself and his batman.

The Digger officer was never a man to dramatise a situation or indulge in histrionics. Even when he was dying he could crack a joke and grin; he never whimpered, never broke.

There was the classic case of Lieutenant Dean of the 14th Battalion, sent out on a bombing raid in France. Back at Company Headquarters they saw two flares fired—Dean's signal that he had captured his ground. Soon after this, Dean, his face covered in blood, staggered into headquarters, already littered with wounded men.

'Here we are, here we are again!' he sang.

He was dangerously wounded, a bullet having lifted the top of his skull, but he refused help until the wounds of the men were dressed. Perhaps he knew he was dying and that dressing would be useless. Either way, it was the act of a brave man. He died there, on the floor of Company Headquarters, and though in great pain he never once complained. He was setting an example.

During the Great War, because of heavy casualties, some officers became battalion commanders in their twenties. At the age of twenty-four, Lieutenant-Colonel A. S. Allen was leading the 45th Battalion at Dernancourt. During World War II he became a major-general. Lieutenant-General Sir Leslie Morshead was in his twenties when he led the 16th Battalion between 1916–19. The youngest battalion commander of World War II was Lieutenant-Colonel (later Brigadier) Sandover, DSO, aged thirty, of the 2/11th Battalion. Officers were so young and so efficient that at a memorial service to two infantry captains killed on the Hindenburg Line, a padre said: 'Such boys to be such men.'

Australian officers and those of some other countries have

a strikingly different approach to fighting. This is best shown by example. Here is the report of an American war correspondent concerning an American engagement in the Pacific.

Captain Leo Arpine's assignment was to take the clump of Jap machine-gun nests on the right flank so that . . . his battalion could drive through this gap and capture the village.

Before the action, Arpine told Love Company, 'This won't be a picnic and we're not going to act like it is. We're going in there to hit these Nips to hell. I want every man up with every other man. Before we go in the artillery and mortar boys will work the Nips to a jelly, so we'll take them okay.'

The support units plastered the jungle with everything they could throw; after a half-hour there wasn't a vine left hanging or a tree that wasn't cut half through. Bazooka rockets, heavy bombs, explosive shells and flame throwers—the Nips got the lot.

Then Captain Arpine gave the signal and they moved in, 100 men in two lines. The stuff we'd thrown at them made the Japs fighting mad; they opened up with small arms and some of Love Company dropped.

Captain Arpine and his squad leaders drove the men on and they began to overrun the Nips. Pretty soon every man was in a personal fight. The Captain had sent some of his men to the flanks to hold back Jap reinforcements. . . .

The fight lasted thirty minutes. The only man on his feet was Captain Arpine; forty of his company were dead and fifty-nine wounded.

'By God!' he intoned to the captain of the support company, 'I said I'd take this position and I did!'

Here is the report of an Australian correspondent about an almost identical Australian action in New Guinea.

Palmer was told to use his platoon to push an enemy force of about forty men from their position across the track junction. . . . He declined a mortar barrage to soften the defences because he said it would only warn the Japs of an impending attack.

Immediately after dark Palmer sent his platoon sergeant and three men through the Jap lines to lie in wait until dawn. At a given time they were to fire on the Japs from the left rear. Palmer told them to stay hidden and to fade away if the Japs closely threatened them.

Under cover of darkness, Palmer, with the three platoon Bren-gunners and four picked riflemen, took up a position to the left front of the Japs' position and less than twenty yards [eighteen metres] away. A corporal held the rest of the platoon in support, for an emergency.

As dawn broke the sergeant's party staged their demonstration. The Japs retaliated immediately. Several exposed themselves, confident that they could not be seen from the front. ... Palmer and his men carefully noted the position of most of the enemy. Then, when ten Japs left their fox-holes to attack the sergeant, Palmer gave the order to fire.

All ten Japs were killed instantly, as were most of the others. The few who survived the first attack panicked when they realised that they were trapped and were shot as they ran. The fight was over in five minutes, without any casualties among the Australians.

In the experience of most Australians who saw the Americans in action in World War II (and admittedly they saw them only in the jungle), about the only manoeuvre the Americans knew was a frontal attack following an extremely heavy barrage. They had every conceivable type of firepower and mechanical device, and they seemed to believe that all-out pressure was enough to get them through. It was, in the end, but they lost hundreds of thousands of men reaching the end.

British tactics have also attracted criticism from Australian soldiers. An attack made by the Scots Guards on a German position in Tunisia in April 1943 is a classic case of how not to conduct a battalion action; the tactics used might have come straight out of World War I. This is how an English historian reported the action:

The Germans entrenched in prepared positions on the hillsides in front of Tunis must have thought the British had gone crazy when they saw three lines of infantry in extended order advancing towards them in broad daylight, as if somebody had forgotten there was a war on and ordered a ceremonial parade.

The Guards had been sent to make an advance of over two miles [three kilometres] under the open sights of the enemy's guns, to scale a boulder-strewn hillside entangled with brambles like barbed wire and to assault a dominant hill at the end of it. ...

The march across the plain was through standing corn, not high enough to give the tall Guardsmen any cover; so that when men began to fall under the murderous mortar and machine-gun fire the only way they could mark where the wounded lay was by sticking their rifles, bayonet first, into the ground. ...

That week had begun when he (the company commander) had marched his men under heavy fire to reinforce another unit, which, when he got there, politely intimated that they were in no need of assistance. Two days later he had lost thirty men in a hazardous attack on a German position which he subsequently held for three days; a bare hillock without shade from the scorching sun or the cascade of shells and mortar bombs. B Company had been tortured by thirst, tormented by flies and shocked by three days of continuous gunfire.

The fumes and dust and the constant cry of 'Stretcherbearer!' had strained the men's nerves to breaking-point. They had been shelled and dive-bombed and machinegunned. Yet now, steady as if they were on parade, the Scots Guards were marching into the mouth of hell again. ...

The Guards lost a lot of men, and the German machine-gun and an .88 mm gun holding up the advance were captured only through the great gallantry of Lieutenant (temporary Captain) Lord Lyell, who won a VC posthumously, for his singlehanded feat of bravery. Lord Lyell might have lived to receive his VC had he been armed with a sub-machine-gun rather than a cane when he led an attack on a German machine-gun post.

The Digger has never seen any military virtue in 'marching into the mouth of hell' or in the acquisition of battle honours to embellish the regimental colours. The infantryman Digger has rarely placed much reliance on armour and artillery. He expects success to follow the taking of calculated risks, his own and that of his leaders.

More often than not the AIF has been handicapped by British leadership. General Blamey, writing to the Prime Minister, Robert Menzies, in March 1941, said: 'Past experience has taught me to look with some misgiving on a situation where British leaders have control of considerable bodies of first-class Dominion troops while Dominion commanders are excluded from all responsibility in control, planning and policy.' Some of Blamey's 'past experience' included the Western Front during the Great War, when he and other Australian officers had seen Australians committed to futile actions, very often against the advice and wishes of unit commanders.

The attitude of senior British professional military leaders and that of their Australian counterparts—few of whom have been career-professional—is different. The British general has, until recent times, seen his troops as inevitably expendable. During the Falklands War of 1982 the army high command was prepared for a loss of 5000 men—though in the end it suffered fewer than 500. The Australian general considers that heavy casualties reflect on his leadership. Beyond this are the respective attitudes of the Australian and British public towards casualties.

For centuries the British have been accustomed to military casualties; they occurred year by year in many hundreds of wars and campaigns all over the world. Military disasters have rarely led to public outcry. The attitude of the Australian public—and government—is just the opposite.

Early in the Great War many Australian senior regimental officers were moulded on British lines and still clung to the rights expected by a British officer. During tough training marches in Egypt they rode on horses while the Diggers walked; this did not endear officers to men. Some did not want to ride but orders laid it down that they should. One training march in particular caused much ill-feeling.

The new 4th and 5th Divisions of the AIF were involved

in the march, made while the Gallipoli divisions with their reinforcements were heading for France. The 4th and 5th were ordered to march from their desert camp to Serapeum on the Suez Canal, seventy kilometres away. It was the toughest march the AIF had faced to that time and by the end of the third day many of them had dropped by the wayside, convinced that the British armies which had served in Egypt thirty and forty years earlier were hard men indeed.

The men were ordered not to touch water, but the sight of a sweet-water canal was too much for them and a bunch of them rushed it, despite the officers who tried to stop them. One soldier shouted: 'They can't stop the lot of us.'

This may not have been good discipline, but then the order forbidding water was not very sensible, since water in moderation is not harmful on a march. It is only when a man takes a gutful that trouble starts—and he never wants a gutful if he is allowed small regular drinks.

General Monash, also on horseback, told some stragglers that their sisters could do better. The troops swore at him. English troops would no doubt have strictly obeyed the order not to touch water, but then it would never occur to them to question the order. Incidents like the canal water rush showed perceptive Australian leaders that Australian soldiers needed special handling and that British Army standards simply would not do. For intelligent men like Monash it was a lesson from which he and others were not slow to profit.

The so-called indiscipline of the Digger has been caricatured out of all proportion. Old soldiers themselves were largely to blame; they told exaggerated stories of doubtful escapades and disobedience of orders, conveniently forgetting the serious side of Australian army life.

An undisciplined army cannot win battles; the AIF won its battles. The British *Official History* of World War I operations in France, noting that in the action of Hamel the Australians had shown how to conduct a large offensive operation, says:

Always formidable adversaries, they [the Australians] had continually improved in their methods and in 1918 had been invariably successful. The staff work of the Corps and its divisions was incontestably of the highest class, and the tasks set the troops were not only carried out by all ranks with innate military skill tempered with training and experience, with reasoned audacity and peerless courage, but also with marked individual initiative and independence and the certainty that, whatever the odds, no enemy could resist them. Of their methods there are no better examples than the St Quentin and Montbrehain, where a few hundred Australians drove the enemy of two, even three German divisions from their entrenched positions.

Undisciplined troops could not do this. The 'undisciplined' reputation came from reports by journalists and British officers who regarded any display of independence as indiscipline. The Diggers were simply different.

Even before they met him, British commanders regarded the Digger as a problem child. General Wavell made himself unpopular with the Australians when he first addressed them in February 1940. With the appalling lack of tact that only the cultured Englishman is capable of, he recalled the Australians' reputation for lack of discipline and told the Diggers that the Egyptians had

> lively apprehensions of what Australians might do in Cairo and elsewhere in Egypt. I look to you to show them that their notions of Australians as rough, wild undisciplined people given to strong drink are incorrect. The Egyptians, generally speaking, are a kind and peaceable people, very easy to get on with. They have good manners themselves and very much appreciate good manners.

Wavell's speech was the type of immoderate address often made to Australians. He had obviously forgotten that he was speaking to intelligent volunteer soldiers, not conscripts. Fortunately for him, the Diggers' irritation was softened by their amusement at his description of the Egyptians. 'Kind and peaceable . . . good manners . . . easy to get on with.' Wavell's rank gave him contact with a class of Egyptians the Australians never met.

The AIF was always highly, if unconventionally, disci-

plined. This discipline came very largely from mateship—
the unwritten creed that said a man could never let down
his mates; that is, every other man in his unit. This is
discipline of the most powerful kind for it breeds an *esprit de
corps* that can never come from worship of the regimental
colours.

It is gross libel to call the Digger undisciplined, to say that
he is a drunk and a larrikin and a thief. He gets drunk
(though I knew many soldiers who were teetotallers), he
sometimes indulges in high spirits and he has been known
to take things which don't belong to him. If he needs food
and ammunition and cannot get it legally he will take it; and
where the opportunity presents itself he will souvenir. He
paints the town red occasionally because he is naturally high-
spirited.

There has only once been a case of Diggers getting seriously out of hand and even that incident could have been
prevented had the English staff officers understood something
of the Digger temperament.

After the Armistice in 1918 the Anzac Mounted Corps
was kept in Egypt and were camped near the village of
Surafend. Relations between the Anzacs and the Arabs, never
very good, soon became strained to breaking-point. The
Arabs, always cunning, had learned that there was a dispo-
sition in the British Army to assume without justification
that any looting and other similar offences practised by the
troops against the natives were committed by the Australians.
If the Arabs missed a sheep they were emphatic that a 'soldier
in a big hat' had been seen in the area.

Australians were disciplined and warned against any type
of scrounging and at this time there was very little of it. But
on the other hand, the Arabs of Surafend and Bedouins from
a nomad encampment nearby were notorious thieves. Night
after night the Anzacs lost gear from their tents. The Anzacs
asked for retaliation, but the British staff were deaf to claims
for justice. Arabs had already shot a few men after the
Armistice and the Diggers were angry.

Confident that they would not be punished, the Arabs
became audacious. One night a New Zealander was disturbed

by an Arab pulling at a bag the Kiwi was using as a pillow. He shouted the alarm to the camp pickets and chased the Arab over the sands. The Arab turned and shot the New Zealander through the stomach at point-blank range; he died as the pickets reached him.

The whole Anzac camp was aroused. The New Zealanders followed the tracks of the murderer to Surafend and put a cordon around the village. In the morning they demanded that the killer be surrendered to them. The head men of the village were evasive and with calm and insolence, pleaded total ignorance; no man from Surafend could have committed the crime. The Kiwis pointed to the tracks, but the head man smiled blandly. Senior staff officers took up the case, but when night fell nothing had been done and the Anzacs knew from experience that nothing would be done.

All day they had allowed nobody to leave Surafend. They were angry and bitter at the death of the New Zealander. A mate had been murdered and they wanted satisfaction, and though nobody has ever condoned what the Anzacs did that night, there were many mitigating circumstances. After dark almost the entire Anzac Mounted Corps surrounded the village. The New Zealanders played the major part, but they were wholeheartedly supported by the Australians. Entering the village, the New Zealanders carefully passed out beyond the cordon all the women and children. Then, with big sticks, they went to work on the Arab men. Several of the Arabs were killed and few escaped injury of some kind. Then the Anzacs burned the village to the ground. As the flames lit up the countryside the Anzacs raided and burned the neighbouring nomad camp. After that they quietly went back to their own quarters.

Early in the morning the full disciplinary machinery of the British staff got busy. The officers who would do nothing to meet the Anzacs' claim for justice now were very busy trying to exact justice for the Arabs.

General Allenby paraded the division in a hollow square and demanded the ringleaders. He should have known better than to ask. When it was clear that he could not charge any individual soldier or even a group of soldiers, Allenby abused the Anzacs, using, as Sir Henry Gullett said, 'many terms which ill became his high position'. The Anzacs would have taken punishment, but they deeply resented personal abuse from the Commander-in-Chief.

The Surafend affair may have had something to do with the dominant role of the Anzacs in putting down the Egyptian rebellion of 1919. The reputation of the Anzacs, already high, was made positively fearsome by Surafend and the rebellion was over in a month.

23

A Gift for Language

Digger adaptability extended to speech and to such an extent that many soldier words are now forever part of the Australian language. Digger language began during the Boer War, but few expressions were preserved. However, the Australian soldiers of those times, called Cockyolly Birds, Gumsuckers, Contingenters, Tommy Cornstalks and Bushmen, could swear with the best of them, according to this comment from the *Bulletin* of 7 July 1900:

> Correspondents in South Africa pay a unanimous tribute to the great Australian Blanky and state that, in curse-language, the man from that great blank continent is laps ahead of Tommy Atkins. When an occasion arrives for extra special profanity, the Cornstalk or Gumsucker is deputed to meet the case and he never fails. Even mules and bullocks which have become absolutely impervious to the indigenous curse wake up suddenly when the Australian attacks with his exotic objurgation.

A war correspondent, Bennet Burleigh, writing in the London *Daily Telegraph* in 1900, also paid tribute to colonial invective:

> ... The very dumb brutes acknowledge their giftedness, for oxen and mules, which would not strain a pound or budge an inch for native or British objurgation, the instant the Colonial takes up his parable, hasten to break thews, muscle and bones, rather than stand stuck in a drift and have such abuse showered upon them.

A South African journalist, who had apparently discovered the Australians and was intrigued with them, wrote that:

they talk in a most colourful fashion, using many words I have never heard. I was under the impression that most of these oddities were oaths and curses, but inquiry revealed that they were, for the most part, slang indigenous to Australia. However, when the Australian horseman does swear it is enough to sear the ears of any sensitive person, though it must be said by the present writer that he has not heard the Australians swear habitually.

During the Great War swearing worried some padres so much that one of them established the Clean Lip Brigade; every member of this remarkable unit was supposed to count ten whenever he felt like swearing and his comrades were urged to pray for him. It was a great chance for the army to issue medals for particularly valorous Clean Lips, but nothing was done. (The British Army in India struck many fine-looking medals for temperance, including one for Five Years' Fidelity. The obverse carried a representation of St George slaying the dragon, while on the reverse was an upstanding bayonet entwined with the words 'Watch and Be Sober'.)

Occasionally during World War II people wrote to the newspapers complaining about soldiers' language and a move was made to revive the Clean Lip Brigade.

However, I am not concerned here so much with swearing, which is referred to in another chapter, as with Digger slang and idiom, which came into being to fit certain circumstances and conditions. Idiom dies once these provoking conditions cease to exist and many colourful slang expressions did not survive the various wars.

Much Digger talk was derived from corrupted snippets of foreign languages—mainly Arabic, French and German—but words were cribbed and adapted from Italian, Russian, Greek and Yiddish, and some from expressions used in the British Army. Many words and phrases were pure original invention, such as *Anzac button* (nail used in lieu of a trouser button) and *Anzac stew* (an urn of hot water and one bacon rind), *drongo* (a useless person), *spine-drill* (loafing).

Many people are under the impression that the Digger's verbal inventiveness begins and ends with vulgarity and foulness, but this is far from the truth. All the vulgarisms were thought of long before the Digger was born and he can hardly be said to have invented any of them.

Digger language resulted from two main causes, the need for lively self-expression and the desire to make light of something serious. The Digger is both an individual and a type and he wanted to speak in a way that would stamp him as someone different. At the same time he has always wanted to conceal his real feelings, to hide completely any trace of sentimentality or emotion. I have known Diggers too embarrassed even to talk of 'my mate'; they prefer instead the cockney term 'my china' (china plate).

Never a formal type of man, the Digger sought to stress his informality and casual nature by a contrived and easy language. In the same way he has never wanted to be known (or branded) as an intellectual; this probably accounts, for instance, for the expression *hairy-belly* for a sycophant.

The Great War produced an enormous crop of Digger words and phrases and among the most durable were *furphy* (a rumour), *maggoty* (angry), *in a pig's ear!* (a contemptuous ejaculation), *possie* (position), *sin-shifter* (padre), *stoush-up* (fight), *throw a seven* (to die), and *upter* (up to putty). A word in common use is *buckshee,* which now means free or gratis, but used to refer to a lance-corporal, who in the Great War drew the same pay as a private.

Other colourful Great War words include:

Arsapeek: upside down

Blow to fook: blow to pieces

Brasso king: an officer who is keen about spit and polish

Buggered: exhausted (also common in World War II)

Bumbrusher: a batman, sometimes known as a dingbat

Camel dung: Egyptian cigarette

Consumption stick: any cigarette

Dekko: a look-see

FIA: forced into action—a play on AIF

Gezumpher: a large-calibre high explosive shell

Junker: superior staff officer

King o' the nits: MP sergeant

Kybosh: the finish ('The sarge put the kybosh on that idea')

Nose-bleeds: red collar tabs worn by senior officers
Pig-stabber: a bayonet
Pot-hole: small trench
To pull on: to undertake a task
Rammies: trousers
Rissole king: army cook, known also as a babbling brook,
     babbler or a hash artist
Shivoo: party
Shickered: drunk
To stonker: do over
Smudged: killed by a shellburst
Snaky: angry
Smoke-stick: rifle
Snottered: killed
Spud-barber: soldier on cookhouse fatigue
Tinkle-tinkle: an effeminate man
Tray bum: very good, from the French *'tres bien'*
Work a passage: to contrive to be sent home
Would-to-Godder: a civilian known to be fond of saying 'I
     would to God I could go to the war'

The much-used expression, 'Dear Bill, ain't it a bastard?'
began its career on Gallipoli. An officer, censoring letters,
came across the significant query at the end of a long tale of
woe from a Digger to a friend at home. Sympathising with
the sentiments expressed, the officer passed on the quotation.
'Dear Bill, ain't it a bastard?' became a verbal sigh of intense
resignation. I often heard soldiers say, simply, 'Dear Bill',
with an extraordinary amount of pathos.

On the Kokoda Trail in October 1942 I came across a
wounded Digger sitting by the track. He had copped one
bullet in the arm and another in his thigh; his uniform was
in tatters and mud was smeared all over him, even on the
bandages which bound his wounds. He was hurt, shaken,
hungry and tired. I lit a sodden cigarette for him and put it
between his lips, but after only one puff it dropped into the
slimy mud. 'Got another?' he said. I had to tell him it was
the last: a non-smoker myself, I had found it in an abandoned
haversack.

The Digger looked down at the mud-covered cigarette, winced with pain as he moved his arm and shook his head slowly. 'Dear Bill,' he murmured. 'Dear Bill.' In no other way could he have expressed his misery and suffering and frustration.

Arabic words used by the Diggers of both world wars include *saida* (good day); *maleesh* (it doesn't matter); *imshi!* (go away); *quies kiteer* (very good or okay); *mungaree* (food); *char* (tea).

Conscripts have exercised the Digger mind many times, resulting in such names as *sugar babies, dingoes, back-to-fronters, wheelbarrows, chocos* and *gutless wonders* (a name also given to the Lockheed Lightning aircraft because of its unusual hollowed-out shape), *stay-at-homes* and *counterfeit-conshies* (conscientious objectors).

A favourite Digger word was *wog,* but it had two distinct meanings. Firstly it refers to a germ, but in this connexion it was of Australian civil origin, not a Digger invention. Secondly, it refers to an Arab, usually in a derogatory way. In *The Australian Language* S. J. Baker notes: 'The Digger use of the word wog for an Arab is doubtless influenced by these meanings—perennial hordes of wogs comprising everything that crawls, hops or flies—and the old nautical wog, a lower-class (Hindoo) shipping clerk.' Many Diggers who served in the Middle East were under the impression that W.O.G. referred to 'Worthy Oriental Gentleman', shortened by the Diggers to Wog.

The Second World War produced yet another crop of words and many have now taken their place in Australian language. One of the most prominent is the exclamation '*Wouldn't it!*'—with an exclamation mark, not a query as some records show. '*Wouldn't it!*' was originally short for 'Wouldn't it root you!' but other versions are 'Wouldn't it tear you!'—'Wouldn't it rip you!'—'Wouldn't it make you mad!'

Other Second World War words which proved durable: *kai* (New Guinea for food); *troppo* (half silly); *doover* (thingumebob); *to come the raw prawn* (to deceive or take down somebody); *bastardry* (anything unpleasant); *a bludge* (soft job);

to dice (throw away); *a galah* (a simple-minded individual); *munga* (food); *no-hoper* (a useless person); *to tee-up* (make arrangements).

Many more words had various lengths of life:

Ack-willy: AWL

Animal: term of contempt for a man

Armchair commando: desk soldier

Attaboys: atebrin pills for malaria prevention

Blot: the posterior

To blow or to blow through: to go AWL

Bombo: cheap wine

Bomb-happy: said of a man suffering from shell-shock or battle fatigue

Bumper-sniping: picking up cigarette butts as a fatigue duty

Bush artillery: captured guns manned by amateur artillerymen at Tobruk

Castor: excellent. For instance, 'She's castor', in the sense of 'She's apples'

Charley the Bastard: the Boyes tank attack rifle, which had terrific recoil

Dead-meat tickets: identity discs

Done over: wounded, injured or exhausted

Emu parade: general camp clean up

February: the CO, because he can inflict a penalty of twenty-eight days in the guardhouse

To fartarse about: to waste time

Free chewing-gum: hat chinstrap

Fruit salad: service ribbons

To front the bull: to be paraded before the CO

To get your finger out: stop loafing and get busy

Get off my back: leave me alone, don't bother me

Gigglesuit: army fatigue suit, so called because Diggers said it looked like an outfit for asylum patients

Give us a break: give me a chance

To give the game away: to be fed up with army life

Homer: a wound serious enough to bring about repatriation

Jungle juice: a home-made and heady jungle drink

Lurk men: loafers and bludgers with a scheme for avoiding work

Mad minute or *mad mile:* bayonet assault course

Mandrakes: waterproof capes, from 'Mandrake the Magician,' a comic-strip character of the time who wore a cape

On the nose: disagreeable or unpleasant

Panic hat: steel helmet

Panic artist: an excitable soldier; usually refers to a somewhat unstable officer or NCO

Pisspocket: a crawler or insincere flatterer

Pull your head in: shut up, you're talking nonsense (This was often used in connexion with troop trains: 'Pull your head in, it looks like a cattle truck')

Perve-man: Over-sexed soldier

Rose bowl: a urinal placed between tent lines. Diggers remarked that this was a good place for men to get their heads together

Short-arm parade: medical inspection to discover men with venereal disease (the 'short-arm' was the penis)

Sharkbait: an unpopular officer or NCO who might be thrown overboard

Stick-it! A contemptuous ejaculation of refusal

The two-icky: the 2 i/c or second-in-command

The trump: the CO

Up shit creek: in a lot of trouble

Woodpecker: a type of Japanese gun

Zebra: NCO (from the stripes on his arm)

Food has always exercised the Digger mind. As far back as the Boer War porridge was *burgoo* to Australian soldiers, though the word is of Old English origin. Sausages were *snorkers* or *snags* and a slice of toast was a *yunk* or *hunk of hard dodger.* Tinned peaches were *Madame Melbas* and the custard was *jerk.* Bullybeef was *bullamakau* and the unappetising army biscuit, an *Anzac wafer* in the Great War, became a *tile* in the next war. In some units *cloud and airships* referred to mashed potato and sausages. During the Second World War powdered eggs brought forth many uncomplimentary terms; perhaps the best known was *chook chuff.*

Oddly enough, neither of the world wars brought to light many words for women not already well established in the

Australian language, except perhaps for *two-bob* (a blonde) and *chicky* (girlfriend) during the Second World War. The Digger has been content to use *sheila, a good sort, broad, bush, brush, clue, dame, doll, tart* and *tab*. Some of these words are American; all are largely without any disrespectful intent. But the prostitute called forth a variety of quite disrespectful terms, including *chromo, fork, bike, unit* (or *camp*) *girl*.

Quite apart from all the words he invented or adapted to his special needs, the Digger has always used much more Australian slang abroad than he would at home—slang which had its beginnings among convicts, squatters, prospectors, sundowners and gold-diggers. This was inevitably part of his aggressive nationalism, his deep desire not only to be an Australian, but to be *seen* to be an Australian. Digger language was a way of emphasising this nationalistic spirit, of under-lining the fact that Australians were different. The Digger doesn't like to be 'treated with ignore', and one way or another he has always made sure that he hasn't been.

During the Syrian campaign a lieutenant of the 25th Brigade was about to launch an attack on a Vichy French post (subsequently found to be manned by the Foreign Legion) but thought he might save lives if he first of all asked the French if they wanted to surrender without a fight. So he sent a captured Frenchman to the post with a message.

The Legion officer in charge read the message and shouted across the narrow no-man's-land: 'We will never surrender to the English!'

The Digger lieutenant lost his block. 'English be fucked!' he yelled back. 'Listen, you pie-eyed, slab-faced bastard, I've got a company [actually he had a platoon] of Australians here and if you don't get out of there within two minutes I'll bring them in with bayonets and have your bloody entrails for necklaces!'

The Légionnaires were surrendering within a minute. It's very difficult to ignore the Digger when he gets his wool off.

24

A Natural Aptitude for Scrounging

As a scrounger, the Digger has always been superb. To be a good soldier a man must also be a good scrounger, a view which may draw fire from some quarters. A British officer once complained to me that the 'Australians are a bunch of thieving scoundrels', but he smiled as he said it. I thanked him for what I regarded as a compliment.

Perhaps it would be wise to define 'scrounging'. Some people refer to it rather loosely as souveniring. But it goes much further than that. It is the acquisition of anything to which the scrounger feels he is morally entitled. The word 'morally' is important. Hence, it is not wrong but is, in fact, commendable.

The Australian, soldier or not, has always been a scrounger. On the goldfields he had to scrounge for all those things that might make his rough home a little more comfortable and his arduous profession a little easier—a candle, pieces of wood for flooring and for making furniture, bits of wire, old wheels and tins and lengths of rope. On the stations and farmsteads he scrounged for the same reason. Australia never gave up anything of its own free will and so the Australian scrounged to live.

Australians were busy scrounging during the Boer War.

When in camp, which is rarely enough [an English war correspondent noted], the Australians manage to live far more comfortably than other troops. They have a natural aptitude to make the best of things and are remarkably clever in adapting the oddest articles for great convenience. and comfort. Not only this, but they seem to be able to acquire equipment not on issue. For instance, I saw a troop of Queenslanders with a regular household stove in op-

Ocker (*Australian War Memorial*)

Blue (*Australian War Memorial*)

A line of walking wounded on their way to a main dressing station in New Guinea, 1942. (*Australian War Memorial*)

These three wounded Diggers in New Guinea are walking back to a Casualty Clearing Station for further medical aid. Their labels describe their treatment so far. (*Australian War Memorial*)

Major General George Vasey, commander of the Australian troops in the latter part of the Kokoda Trail campaign, 1942–43. (*Australian War Memorial*)

eration and some men from New South Wales had a cart in which they took along extra items of personal equipment and the souvenirs all the Australians seem to regard so highly.

A Victorian captain was said to have scrounged some tents—though the correspondent used the word 'appropriated'—because his men did not have enough. A subaltern was ordered to return a captured Boer Maxim which he had also appropriated for additional fire-power.

A Rhodesian historian noted that the Australians were 'marvellously proficient at making things from pieces of wire and tin—hangers, eating implements, candle-holders, plates and all manner of things'. 'More often than not,' the same writer relates, 'many of the things in an Australian camp have been taken from the Boers themselves. The Australians never throw away anything.'

Military scrounging during the Great War had its roots in necessity. Troops had poor clothing and equipment; and never enough of it. If they could scrounge something better, then they did. They cased the bodies of dead enemies—and dead friends—not only to souvenir watches and money and fountain-pens, but to get themselves a better pair of boots and a warmer coat and some more ammunition.

The Digger, ever a realist, sees no harm in robbing the dead; for obviously a dead man cannot read a watch or spend his money. Because he himself still lives the Digger sees himself entitled to what the dead can no longer use. I would be the last to dispute it, for I have done my share of scrounging.

Sheer military necessity has resulted in some spectacular scrounging. A young officer ran out of ammunition for his platoon and was told that he had drawn his issue and no more was available. But he knew very well that a British ammunition park, some distance away, had more cases of .303 ammunition than it could find room for. He commandeered a truck and with his sergeant and two men drove to the park where he bluffed the guards into admitting him. The English captain in command said that he could part with no ammunition without authority.

'That's all right with me,' the lieutenant said, uncon-cernedly. 'I'll tell General Blamey what you said.'

'General Blamey?' the English captain repeated, startled. 'Why does *he* want ammunition?'

'I'm only a lieutenant,' said the Australian. 'I don't cross-examine generals. I'll tell him you want him to sign some forms.'

He began to climb back into his truck, but the English officer stopped him. 'If you'll sign for the ammunition . . .' he said.

The Australians loaded up, the officer signed his name 'Lieutenant Ned Kelly', and the truck drove off. 'Ned Kelly' felt no particular elation; he needed bullets and he got them, that was all.

Ned Kelly was responsible for an extraordinary number of misdeeds. An English colonel once asked me if Kelly was a common Australian name. The poor devil ran a base clothing depot and the number of times Ned, Edward and N. Kelly took him for unauthorised clothing was just a shame.

On Gallipoli the Anzacs became clever and enthusiastic scroungers. E. J. Rule tells of two quartermasters who were fine scroungers.

> With eyes like hawks, they spied out everything that was moveable and not nailed down, and, just as soon as dusk began to fall, the dirty work began. Our QM (Larter) invariably asked me to go along and I rather liked it. On this particular Sabbath night we were out to 'lift' some water drums down near the beach. In the long sap the only signs of life were the Indians and their mules. Long before we could see them in the gloom we could hear the steady beat of the animals' feet and the scraping of their loads on the side of the trench. As soon as they came in sight our QM would stand still on one side of the sap and commence calling out, 'Good night, Johnnie', until the last one had passed. Larter put it this way: 'Frighten any of these fellows in this lonely bloody sap and you might find their kukri hanging in your ribs.'
>
> When we saw what Larter wanted us to carry back we demurred. They were the iron water-containers in which

the mules brought our daily rations. Larter offered me a quarter of a side of stinking bacon (good for making candles) if I could get mine home. In the morning I went round to claim my reward and I found the QM peeved. 'Here's your bacon,' he said, 'but didn't you see these blasted bullet holes in the drums?'

Some scroungers of all Australia's wars became so addicted to the habit that they could not resist the temptation to take anything left lying around. One infantry company in France had a member called by his mates 'Klep'—for obvious reasons. His depredations became so active that at last his long-suffering mates complained and the OC heard of it. 'Klep' was called to the company office and stood at attention for a ten-minute lecture.

When he came out of the office one of his mates said, 'Well, Klep, what did you pinch this time?'

'Oh, the bugger didn't have much on his table,' complained 'Klep', 'but I fiddled this while he talked!' From his pocket he took a well-filled inkpot.

Scroungers must take the bad with the good. In Port Moresby in 1942 a friend and I raided an RAAF sergeants' mess during a Japanese air raid, when the airmen were in their shelter. They lived well and we scrounged a heavy supply of unlabelled tins; we guessed that they would contain apricots or peaches. A couple of weary kilometres later we sat on a hilltop, opened the tins with our bayonets and found carrots.

On Gallipoli scrounging became a fine art, because there was not a great deal of scope. To frisk a body left in the field for weeks was an objectionable job. It was the duty of anybody—and particularly an officer or NCO—to take from a body one of its identity discs for record purposes and as duty compelled him to do this much, many an NCO felt himself entitled to payment for the job.

Rule notes that in France

Every body that lay out was sooner or later ratted [searched and their effects taken] either in the course of duty by AMC men, whose business it was to send the effects into their headquarters, or by passing soldiers who were hunting for what they could find. Almost invariably a man's haversack would be opened, anything of value taken out of it, and the rest scattered around his body. Where a man had been dead some time the pockets would be ripped open with a knife.

In the end during the Great War, nobody owned anything, for army life changed a man's conception of right and wrong. Stealing came to be accepted merely as borrowing. Many officers said flatly that they would recognise no shortages.

In March 1917 a plane piloted by Prince Charles of Prussia crash-landed several hundred metres in front of the Australian posts near Lagnicourt. The prince clambered out and ran down a valley. Corporal B. G. James of Queensland and Corporal E. J. Powell of Tasmania (killed a year later) shot at and wounded him and then ran up to him. A couple of Light Horsemen also galloped up, so the two corporals divided the prince's cap, gloves and goggles with them, as souvenirs. The prince died in hospital a few days later, but not before thanking the Australians for their 'kindness and good sportsmanship'. He, too, was a good sport, he told them.

Diggers went on leave with the idea of scrounging quite as much as they went looking for drink and women. From long association with the Diggers, General Birdwood knew well all their delights and desires. It was widely rumoured in the British Army in 1918 that when the problem came up of capturing the great French city of Lille without a preliminary bombardment—which would destroy the place and cause civilian casualties—Birdwood presented a scheme to GHQ. His idea was to bring an Australian division into the sector; then each man would be given £10 advance of pay and four days' leave. Finally, it would be announced that Lille was out of bounds to all troops. Birdwood guaranteed that 80 per cent of the Australian division would be in the city within twenty-four hours and that the Germans would have the good sense to get out.

Scrounging has its penalties for the careless and over-enthusiastic. The Germans and Italians realised this during World War II when they littered the Western Desert with fountain-pens, vacuum flasks and other novelties rigged as booby-traps. Several Australian soldiers lost their lives or limbs before the attractive souvenir-traps became suspect.

The Digger's habit of kicking tins and anything else lying in his path also led to booby-traps. A tin could hold down the striker lever of a grenade, which would fly off when the tin was kicked away. All the booby-traps set for the Australians of World War II followed the lessons learned by the Germans during World War I. The Digger would always fall for a helmet left on the ground; a helmet had high souvenir value. And often enough it was a booby-trap.

When the Diggers took Péronne in 1918 one of their first jobs was to clear the streets. Two of them found a dead German lying on a stretcher and picked the body up to carry it away. Stretcher, German and Diggers were blown to pieces by the mine attached to the stretcher with a piece of wire. The Diggers on this occasion were not scrounging, merely carrying out normal duties.

An officer lived in a German dugout for three days before it blew up and killed him. Somewhere in the roof was an acid slowly corroding a wire, which, when eaten through, permitted a mine to blow up. Peronne was put out of bounds for two months while engineers deloused the place.

Lift the latch of a door and it blew up in your face; tread on a loose step, open a cupboard, wind a clock—they all worked against the Australians until they learnt better, but nothing could stop them from scrounging.

A colonel walked into a house used by the Germans as a command post and saw a gold watch hanging by a copper wire on the wall. Carefully he attached a string to the wire and paid it out until he reached a shell-hole. After he had given the string seven or eight heavy tugs it came loose, without the watch. He returned to the house and found the watch gone. The whole thing puzzled him for a week. Then he found a Digger with the watch. He had entered the house while the colonel was yanking at the string; he took the watch and left. The colonel wanted that watch, but the

Digger protested loudly, pointing out that he had a moral right to it. The colonel might have seen it first, but who took the greater risk in getting it? The logic of his case appealed to the colonel, and the Digger got the watch.

In the great tunnel through which the Saint-Quentin Canal ran there were found behind an apparently innocent boarding a number of shells with percussion arrangements fixed to the fuse which would be set in action by anybody casually leaning against the boards. The most innocent-looking objects, such as pocket-books, pictures, chairs, helmets and other things exploded when picked up. On two occasions there were found apparent graves marked with a cross and the inscription 'Unknown Englishman'. Suspicious engineers found that the graves contained a mine, not a body. Even in churches delayed action fuses were built to go off at times ranging from twenty-four hours to three weeks.

The Germans often arranged shells with a delayed action fuse of the wire and corrosive liquid type to blow up ammunition dumps. Sleepers on railways had high explosive shells beneath them set to explode under the weight of a train.

Special companies of trained volunteers—they wore a broad stripe of red tape down both sleeves—discovered 14,000 German mines and traps of various descriptions containing 550 tonnes of explosives, and deloused them all without mishap to themselves. Called the Red Tape Merchants, these men from the Royal Engineers followed hard in the wake of advancing infantry and prevented them from killing themselves with booby-traps. They carried with them field postcards and when an engineer had deloused a trap, shell or mine he would write on the card 'Examined—made safe', sign it, leave the card with the trap and go his way. These anonymous men saved the lives of many Australians, who by their very nature were more prone to fall for booby-traps than soldiers of the other Allied armies.

Even at the most desperate moments Australians have lingered long enough to scrounge a few souvenirs. Sometimes they have lingered too long. Two were captured in Greece because they dallied to souvenir a set of Greek coins.

Enemy troops of several nations and wars have been astonished at the rapidity with which they were frisked. Describing the capture of Germans on the Western Front, E. J. Rule said:

> As each Hun advanced with his hands above his head several of our lads would dive at him, and, before the Hun knew what was happening, hands were in every pocket, and he was fleeced of everything except his name and his clothes. . . . A new reinforcement was working like a cat on a tin roof pulling cigars out of a Hun's pocket.

Company cooks have been among the most expert of scroungers, for the good army cook is dedicated to the men he serves and will go to any lengths to give them good meals. During the Great War some of them developed the art of bandicooting potatoes from French farm-fields. Bandicooting is stealing potatoes without leaving tracks. The cooks would scratch down under a plant, take two or three potatoes, and cover the hole again. The plant didn't die and the next fall of rain would set the ground again. French farmers were puzzled at their small crops.

In Syria in 1941, as elsewhere, the Australians were in trouble. Ten days after hostilities against the Vichy French had ceased General Spears of the British Army wrote to General Auchinleck, the C-in-C:

> The Australians are already greatly feared by the natives. Their behaviour, with the exception of some specialised units which are well disciplined, would be a disgrace to any army. They are alleged to have stolen Vichy officers' wedding-rings and to have deprived prisoners of their water bottles. At Mezze aerodrome by way of contributing their quota to the efficient conduct of the war, they stole and smashed vital parts of the Air France wireless installation.

General Blamey had these allegations investigated by Colonel Rogers, senior liaison officer of the 1st Australian Corps. After a thorough enquiry Rogers reported that the accusations were unfounded, also, in one respect, grossly libellous, mischievous and irresponsible.

Blamey was always quick to defend his men, a fact which few of them appreciated. On 11 October 1941, General Wilson wrote to Blamey complaining about 'brutal assaults by Australians against soldiers [British and French], police [military and civil] or civilians'. Wilson asked for exemplary punishments and suggested that a 'whack of penal servitude with the first two years to be done in the Middle East' might have a deterrent effect. Blamey pointed out the previous accusations against Australians had been proved baseless. 'It is a very convenient form of excuse for any happening to lay it on broad Australian soldiers.' He told Wilson pointedly that the question of discipline in the AIF was entirely one for his action and not Wilson's.

Neither Blamey nor any other officer, senior or junior, condoned violent and irresponsible conduct or thieving, and the men of the AIF knew it. About twenty soldiers were found guilty of conduct which earned them gaol sentences.

It was indeed libellous for Spears, Wilson or anybody else to say that the AIF lacked discipline. While they had none of the servility of the British troops and sometimes looked untidy on leave, they were very rarely guilty of unprovoked assault.

At the signing of the armistice between the British and Vichy French at Ancre on 12 July 1941, somebody souvenired the gold-leaved képi of General Catroux. Catroux believed that an Australian took it, and General Wilson shared this view. In his book, Catroux said:

The camp [where the armistice-signing took place] was occupied by Australians who, as is well known, besides being keen fighters have an instinctive belief in freedom of action. They are keen on bringing back souvenirs to their distant land from their travels or military campaigns. I was imprudent enough to leave in my motor-car my oak-leaf képi, the gold embroidery on which fascinated some son of Australia. It was a fine souvenir to carry off. ... Need I say that the camp commandant was all the more upset because he well knew that I would not get my képi back. I reassured him by dissuading him from a search and said to him that I had commanded fellows very much like his Australians—I meant the Foreign Legion—and knew by experience that what was taken by them was taken for good. The incident ... made Lavarack smile. Lavarack was commander of the Australians and well knew what his soldiers were likely to get up to.

The *Official History* notes that 'only four Australians were present and it is improbable that any of them took General Catroux's képi'. I admire the historian's defence of the four Australians. I happen to know that an Australian did take that képi—and it wasn't one of the four to whom the historian refers. It was a young officer who had occasion to pass the general's car, saw the képi and decided he badly wanted it. 'The general,' he told me, 'could easily get another.'

Food and drink has often been the object of scrounging and many a jar of rum or case of fruit has gone off while the attention of guards has been distracted by a brawl specially staged for that purpose.

The truckloads of Diggers who rolled through the orchard country of Lebanon in 1941 looked longingly at the laden orange trees, but usually there was an NCO riding in the truck cabin and orchard raiding was therefore difficult.

A pair of schemers in the back of one truck hit on a plan of action. Near a bend in a road on which they had travelled previously was a large orchard and they decided that near the gate of the orchard they would toss out a hat. Then, as they passed the bend, they would call to the driver to stop while the hat was retrieved. Since the NCO could not see what happened next they reckoned they would be safe.

The hat was thrown, the truck stopped and the two Diggers scrambled over the side, raced back to the bend and struggled through the fence wire. They were stuck in the fence when an elderly Lebanese came through the orchard towards them. 'Good morning,' he said politely. 'I have three cases of oranges waiting for you at the gate if you come and carry them please.'

The shame-faced Diggers took the oranges—but insisted on paying for them. Still, they weren't out of pocket at the finish; they sold the oranges, cut-rate, to their mates.

At Christmas 1940 some members of the 2/17th Battalion, then in Palestine, used a daring plan to get more beer supplies. After they dressed as guards, a private posing as a sergeant marched them to the huge dump of canteen supplies at Cherriff Quay. There, with the full formality, they relieved the British guard. As soon as the old guard had moved off a truck drove up and the 'guard' loaded cases of beer on to it. Then they buried the beer, mainly in the sandy floor of the tents. Even Brigadier Murray had a Christmas drink of the stolen beer without knowing its source.

Nobody knows how many times a couple of Diggers appointed themselves as guards and reported for 'duty' at a hotel or café, telling the proprietor that they had been sent to prevent disorder in his establishment. Mostly they had no intention of scrounging anything, but relied on the proprietor or manager to provide free and plentiful food and drink, which he always did. I knew a lance-corporal and two privates who worked this dodge in Townsville for nearly a month, filling themselves up each night on beer and rich food supplied by the hotel staff. Then they were really rostered for duty one night and did not turn up at the hotel. The proprietor, who had become used to his guard and even felt that they gave his hotel some insurance against possible brawls, phoned the camp to report that the 'night guard' had not arrived. You might think this was the end of the honeymoon. It was, but only for this particular lance-jack and his mates. Camp Headquarters, assuming that a guard should be posted at the hotel, thereafter detailed a corporal and two men as a standard nightly duty.

When a Digger decided he wanted something he usually got it—one way or another. In 1917 two Diggers sent to collect something from a railway station in France noticed that near the door of the office occupied by the Railway Transport Officer, an English captain, was a pile of rum jars—all full. Unfortunately, the rum and other stores were guarded by two English military policemen. The presence of the Jacks only increased the Diggers' thirst—and determination to get some of that rum.

They collected what they had come for, carried it to the waiting cart, put their heads together and returned to the station. They approached the RTO in his office and one said: 'Could you tell me what time the train for Paris leaves, sir?'

'Two o'clock,' the captain said absently.

They thanked him and departed, but a few moments later one Digger put his head in the door and said: 'You did say two o'clock, sir?'

'Yes,' the RTO said impatiently. 'Two o'clock.' Then he looked up suspiciously. 'Do you have a leave pass for Paris?'

'We're not travelling, sir,' the Digger replied. 'We're worried in case our company commander misses the train.'

'Oh,' the RTO said. 'Well, two o'clock.'

The Digger thanked him and again the two of them stood on the platform and passed the time of day with the Jacks. Had the MPs been alert they might have wondered why two Diggers were being polite to them. After a few minutes one Digger said resignedly to his mate: 'I s'pose we'd better finish the job, Slim.'

'All right,' Slim said. He put his head in the door of the RTO's office once again and said: 'It *was* two you said, sir?'

The RTO lost his block. 'Yes, confound it!' he shouted. 'Two! Two!'

'Thank you, sir,' Slim said and saluted. 'Just two.'

He turned back to his mate and nodded at the rum jars. 'The RTO says two, Harry. I thought the QM said four, but he and the RTO can fight it out between them.'

Then they each picked up a rum jar, nodded to the Jacks, and calmly walked off the railway station.

The men of a Light Horse regiment used to tell a story about 'Johnny' Walker when he was orderly sergeant of the day. Having scrounged a lot of cheap wine while in Cairo he did not remember until about three o'clock in the morning that he was rostered as orderly sergeant. Befuddled, he nevertheless found a gharry driver to take him back to the camp at Ma'adi, some kilometres from Cairo, and arrived just before reveille sounded. 'Johnny' staggered into his tent to prepare for parade. Out tumbled the sleepy regiment for roll call and the men could hardly believe their eyes when they saw their orderly sergeant 'all complete'—with polished leggings, bandolier, and spurs shining. Pulling out his roll book 'Johnny' swayed unsteadily but found a bit more courage and began the roll call.

'C Squadron (hic) answer your names! Walker!' There was no answer but a general laugh.

'Walker, answer your name!' Just then an officer appeared.

'Walker!' shouted 'Johnny'. 'Where'sh that man Walker?'

'Sergeant Travers!' the officer said sharply. 'Escort Sergeant Walker to his tent. He is under arrest for being drunk on duty.'

'Johnny' had tried very hard; it was just his bad luck to call his own name first—and get no answer.

A particular form of scrounging had a definite profit motive. I never saw this better practised than in New Guinea in the Second World War when some enterprising Diggers, fascinated at the amount of money the Yanks had to spend and wanted to spend, decided that they had to get hold of some of it. The Digger was always a shrewd psychologist and he recognised these Yanks for what they were—exhibitionists and braggarts. Some of them paid more that $200 for a Japanese sword so that they could take it back to the States and boast of having taken it, in bloody combat, from a Japanese officer.

This bunch of Diggers started a part-time business—manufacturing Japanese battle-flags. Because of its simple design, the Japanese flag was easy to make and when stained with green dye or red ink it was a most convincing souvenir of some fight in the jungle. The Yanks queued in dozens to pay up to $20 for a flag.

Other Diggers cashed in on the racket by selling Japanese invasion money. It was genuine Japanese-printed money, but worth hardly anything even as a souvenir. Still, the Yanks paid out good dollars for it. When the Diggers found out just how completely gullible the Yanks were they took to firing a bullet through a wad of invasion money; it then brought a higher price because it 'had been taken from the body of a Japanese general by the Australian who shot him'.

Still other Diggers sold Japanese grenades to the Yanks

for as much as $40 apiece. At one time, many of these captured grenades were stored in Port Moresby, so the supply was plentiful. Japanese grenades were notoriously unpredictable and sometimes exploded when looked at (so the Diggers alleged) and the vendor was always quick to move off after closing a deal.

The Australian mobile workshops in New Guinea probably had the best racket of all. With the very best of equipment available they made finger rings from pieces of metal salvaged from crashed aeroplanes and inset them with initials or designs made from the plastic in toothbrush handles. The Red Cross in New Guinea was puzzled at the vast number of toothbrushes constantly being requested by the troops and as far as I know they never did discover where the brushes were going.

A private from a mobile workshop offered me five shillings for half my toothbrush handle, though I believe the rate later went to five shillings an inch, which was good money if you had a lot of toothbrushes. At the beginning, the sales story was that the rings had been taken from dead Japanese, but so many rings flooded the market that even the most innocent GI from Little Rock, Arkansas, must have wondered if every Japanese on service in the Pacific had a ring on each finger, and this selling point was dropped. Anyway, the Yanks were willing enough to buy a ring merely because it had been made from 'the wing of a shot-down Japanese Zero'.

Actually, with very few exceptions, the rings were made from pieces of crashed American Liberators, Douglases or Lockheed Lightnings. From rings the workshop men progressed to brooches of all kinds, the speciality being ornaments with a quasi-Japanese design. Some fantastic sales stories were woven around the origin of these brooches; some were the heirlooms of illustrious Japanese families, others had been worn by geisha girls brought to the Pacific Islands for the comfort of the troops. No matter how fanciful the yarn the Yanks swallowed it—and paid up.

Eventually, some operators were making 'genuine' Japanese daggers, comprising a blade from half an Australian bayonet and a hilt from a piece of wood salvaged from a broken rifle-stock and then polished.

Most of these items went back to the US to be displayed as 'battle trophies' by Yanks who hadn't been within a hundred kilometres of the enemy. While the market lasted the Diggers concerned cleaned up in a big way. It was a form of scrounging that the Diggers of the Great War never did get a chance to practise.

I scrounged many things myself, but I specialised in ammunition and explosives of all kinds. Originally, overseas, this had a sound basis, because I always wanted to have some extra fire-power 'just in case'. Some troops, even though well trained, might lack experience and tend to expend ammunition too quickly. An officer or NCO liked to have a reserve for an emergency such as might arise from too enthusiastic firing by his men. Back in Australia, serving with a training battalion, I used to keep bullets, grenades, mortar bombs and explosives left over after various training exercises. This was strictly against orders but I felt that 'it might just come in handy'. I stowed the stuff in a couple of old ammunition boxes under my bed.

Apart from several hundred rounds of small-arms ammunition, for rifle, submachine-gun, tank-attack rifle and pistol, I had about two dozen '36' grenades (Mills bombs), the same number of HE '69' percussion grenades (they had a bakelite casing), ten 2-inch mortar bombs and one 3-inch mortar bomb (it weighed 4.5 kg), a quantity of explosives including gelignite, monobel, ammonal and guncotton (TNT); enough in all to blow up several railway lines and a couple of bridges.

I eventually took this stuff home and kept it under my bed, until my wife discovered what she was sleeping over and left the house with the threat not to return until I got rid of my box of treasures. I kept some of the SAA and dumped the rest of the collection in the river. A pity—many a time since then it would have come in handy.

Actually, this is what motivated most Digger scrounging— the provoking thought that rope, a box, plate, spare belt, ammunition, wire, tins of food, an enemy weapon, and so on *ad infinitum*, might come in handy. It hurts to throw something away and it hurts more when you find, later, that

you could have used the discarded item. Except gas respira-
tors; no Digger in the Second World War felt any distress
about losing a respirator.

I heard a lieutenant-general address a meeting of officers
this way: 'Gentlemen, there is a distinct difference between
pure [his word] scrounging and thieving. I will not tolerate
the appropriation by troops of property belonging to the
civilian populace or the theft from restaurants and other
public places of glasses, ash-trays and cutlery. You will im-
press upon your commands that a serious view will be taken
of this. However, you will at the same time encourage pure
scrounging because it develops the soldier's initiative and
resourcefulness. Any questions?'

A brigade major, speaking for the group, said: 'Would you
define pure scrounging, sir?'

'Certainly,' said the general. 'I have no objection to any
soldier salvaging any potentially useful material from waste
dumps or picking up and keeping any discarded object, provided its retention does not impede his movements. I have
no objection, when in fixed camp, to troops improving by
scrounging their physical comfort. Nor have I any objection
to troops using their wits to secure items from any source—
even the quartermaster—provided the other party is given
an even break. But *that*, gentlemen, will not appear in routine
orders! If a quartermaster cannot see through ruses thought
up by the troops he should not be a quartermaster. Understand this, gentlemen, I will not have troops punished for
the type of scrounging that reflects enterprise.'

Sometimes it must have been very difficult for the general's
battalion commanders to draw the line between thieving and
pure, enterprising scrounging. There was the CO who had
to adjudicate in the case of a private who was found to have
in his possession three sets of underwear on issue to the
Australian Women's Army Service, and the artillery officer
in Syria who was faced with the problem of two gunners
who had a dressing-table, complete with mirror, in their tent.
The private claimed that the underwear was a 'sentimental
souvenir' and the gunners said that they had found the dressing-
table abandoned in a wrecked enemy fort.

I wonder how the general would have viewed the case of a corporal I knew named Bill Smith (this really was his name). Corporal Smith had a new angle on scrounging. A lurk man, he scrounged nothing more than *freedom within the army*. I first knew Corporal Smith in New Guinea, but before that he had been in the Middle East and had done a lot of soldiering.

He left my unit when it was in the Atherton Tablelands of northern Queensland and I did not see him again until one day in September 1944, when I was walking through the incredible no-man's-land that was General Details Depot at Sydney Showground. He was neatly dressed in khaki shorts and shirt and was still a corporal, but to my surprise he carried in his right hand a large hammer and in his left a bag of carpenter's tools.

'What's the hammer in aid of?' I asked.

'Me,' he said. 'I'm a carpenter.'

'Yes, and I'm an interior decorator!' I said.

'Dinkum,' he said. 'And I'll tell you how I got this way— if you don't put me in.'

We found a seat in the sun and out of the way and Bill Smith told me his story. 'I was browned off and fed up to the back teeth with the war and the army after New Guinea,' he said. 'All I wanted to do was get out and go back to the sheds [he was a shearer] but the bastards wouldn't discharge me. I told them I'd had a gutful and somebody else could have a go, but they said I was experienced and fit and that Blamey needed me. I came into GDD for posting, and hung around for about a week but nothing happened. Every morning I'd report for roll-call, then sit on my backside all day. This joint was full of lurk men, so I figured I could get some sort of cop for myself. I scrounged this tool kit and carried it into the orderly room that had my name on the roll and told a sergeant that I'd been attached to the GDD carpentry unit. He said okay and gave me a transfer to take to the carpentry orderly room. I think I've still got that transfer somewhere.'

I was a bit out of my depth. 'But didn't the carpentry orderly room ask for the transfer?'

Private B. T. McMahon of the 2nd Battalion Royal Australian Regiment on sentry duty in Korea. He is wearing a flak jacket and carrying the then-new British Patchett 9-mm machine carbine. (*Australian War Memorial*)

A badly wounded Digger of the 3rd Battalion Royal Australian Regiment receiving field treatment in Korea in 1951. (*Imperial War Museum*)

Private C. D. ('Dig') Coy resting before going into action in Korea. In the background is a company 'hoochie' or cookhouse. The white tape was used to guide soldiers moving by night. (*Australian War Memorial*)

A sergeant of 'D' Company 8th Battalion Royal Australian Infantry on patrol in Phuoc Tuy province, Vietnam. With no possibility of quick re-supply, he is heavily armed and carries a lot of ammunition. (*Australian War Memorial*)

'I didn't even report to them,' Corporal Smith said. 'I don't even know if there *is* a carpentry unit. I'm a one-man unit. I roam about GDD three days a week doing little jobs. I bang a few nails into a wall of the officers' mess or screw some hooks into a door or tighten a hinge or say I've come to check on the chairs. I take on anything that doesn't mean dinky-di carpentry. Everybody around the place is used to me now, but I'm dead scared to put my hammer down. If I lose that, I'm buggered.'

'What happens on the other four days of the week?' I asked him.

'I'm at home,' Corporal Smith said simply. 'I've got the missus down from the country and we go to the beach or the pictures—anything to kill time. I draw my pay regularly, of course. It's not a good life, but it's not a bad one either. And the war's got to end sometime.'

'But how about leave passes?' I said. 'How do you get out the gate? And what happens if the Jacks pick you up?'

He grinned at me. 'I'm too well trained to try to get out without a leave pass. I scrounged two whole books of leave passes and a rubber stamp. I write my own. I might have to scrounge another book soon; these two are about finished. 'Course, the war might be over before I need any more.'

'You've worked through *two* books of leave passes?'

'Well, I've been working this lark for a year,' he said. Then he went off to drive a few more nails into another wall.

To remain AWL within GDD for a year might sound incredible, but it was, in fact, very simple. Thousands of soldiers moved through the enormous depot every day; rolls changed constantly and officers and NCOs in charge of the various companies to which men were temporarily attached never knew men by sight. The whole showground was full of orderly rooms and sub-units of every kind. Squads of men marched this way and that and many hundreds meandered aimlessly about while awaiting drafts or decisions, because there was no way of keeping them occupied.

At least once during 1944 MPs made an organised raid on the many hideouts in GDD while a muster parade (a com-

pulsory parade for every man in camp) was in progress. The
Jacks rounded up more than 200 soldiers who had been
AWL for varying periods, but living and eating in GDD.
When I met Bill Smith after the war he told me that he was
'on leave' at the time of the raid.

It is obviously impossible to award the biscuit for the most
spectacular feat of scrounging performed by a Digger, since
neither I nor anybody else knows of all the scrounging
performed. Probably there were several thousand 'best
scroungers'. However, I do think that a New Zealand sergeant performed one of the most astonishing feats of scrounging ever, even though it could not be called strictly orthodox
scrounging.

Serving in Italy during World War II, the New Zealander,
impressed with the many ways of making money illegally,
hit on what must be one of the most successful frauds in
military history. His only equipment was a rubber stamp he
had specially made, a pad of ordinary army forms concerning
requisitions and a convincing manner.

He would visit an Italian known to have an expensive car
and tell him, regretfully, that he had orders to commandeer
the vehicle for the Occupation Army and had come to take
it away. The Italian, being polite, would suggest some wine,
the New Zealander would accept and the two would have a
cosy chat. Eventually the Italian would offer a bribe to be
allowed to keep his car. After a show of hesitation, the New
Zealander would accept the bribe 'because you're a genuine
case and I can see it would be pretty tough for you to lose
your car', and would stamp a requisition form EXEMPTED
FROM REQUISITION. Signing the form with a false name and
pocketing the cash bribe, the sergeant would shake hands
with the Italian and depart.

Civilian cars were *not* being requisitioned, so he was safe
from the possibility of somebody later calling on the Italian
with a genuine order of requisition. The Kiwi is supposed to
have made as much as £1000 a car, though the average bribe
paid was about £250.

Of the 440 rackets known to the British Army Investiga-
tion Service this one was the neatest of all—and nothing was

ever proved against the Kiwi. Some Australian ordnance and supply officers were involved in the illegal sale of army refrigerators and other equipment, but none ever showed the complete audacity of the New Zealand sergeant.

25

Swear at It, Laugh at It

Digger humour has always been lusty, earthy and ironic, and some of the best examples are unprintable. This is not to say that Australian soldiers are any more dirty-minded than other troops, for off-colour stories are common to all armies and to any community of men denied the restraining and refining influence of women. The men of a French Army post in Algeria, where I once spent a week, were by far the foulest-mouthed men I ever encountered and many American troops I met told stories that the average Digger would brand as 'plain filth'.

The Digger himself has no liking for plain filth, though he will tell and listen to doubtful stories and enjoy both the telling and the listening. Inevitably, sex and women are the basis of the majority of stories. When a man is denied something he badly wants he will talk about it—and so for centuries soldiers have talked about women.

But Digger humour goes far beyond sex. The Digger will joke about anything—his officers, camp facilities, leave, his mates, equipment, the enemy, even death. His humour is sometimes sharp, mostly ironic, always pertinent. The gag that would make a YMCA hut full of Tommies roar with laughter would barely raise a smile from a Digger. Many an English concert-party performing for the Diggers thought they were unappreciative, but in fact pie-in-the-face slapstick humour has never gone over with Diggers. They like a bite to their humour.

The Digger could always laugh at his own misfortunes and even when he grumbled he was, subconsciously, poking fun at the object of his grouch—even if the object happened to be a senior officer.

On the Western Front in 1917 one morning a brigadier

visited the front-line trenches to impart some cheer. The
men were standing in a metre of mud as he addressed them.
'I have great news for you boys,' he said heartily. 'The French
attacked last night and captured thousands of prisoners.'

Blank, expressionless faces looked up at him. Disappointed
at the lack of response, the brigadier repeated his information.
After another pause one Digger said: 'Do you think we're
winning, sir?'

'Of course we are!' the brigadier said.

'Well, if we're winning,' said the Digger wearily, shifting
his invisible feet in the mud, 'God help the poor bloody
Germans.'

The mud of the Western Front was an insidious enemy
which never retreated. It was so thick and soupy that it
sucked off boots, puttees and trousers. E. J. Rule tells of a
man he saw coming from the front-line, wearing only a tin
hat and shirt. Shivering in the icy wind, he picked his way
along in his bare feet, the tail of his shirt stiffened with
slime. When several watching Diggers laughed, the sufferer
exploded. 'Do you think you're at the flaming zoo?' he
shouted. 'Hop out, any of you bastards, and I'll give you
something to laugh at!'

A British general inspecting the trenches one morning
complimented an Australian captain on his position.

'Yes, it's not a bad possie,' the Australian said.

'Possie?' exclaimed the general. 'What is a possie?'

'It's an Australian word for recess, either firing or sleeping,'
the captain explained. 'It's really a contraction of the word
position.'

'Oh, I say, surely not!' the general's accompanying staff
officer objected. 'Of course you mean, posse. P-O-S-S-E—
meaning a small force. Your firing recess is manned by a
small force, what!'

'Lynch the bastard!' a listening Digger said.

'What was that?' the general said, startled.

The captain smiled appeasingly. 'Just a member of the
posse, sir,' he said.

Any particularly stupid comment or action by an officer
inspires the Digger to biting, satirical humour.

While its own doctor was absent a certain battalion in
New Guinea in 1943 had the services of a stand-in medical
officer, a Scot who had emigrated to Australia. He seemed
to have no other remedy than 'Take two aspirin, mon'. This
was his panacea for everything from a black eye to tropical
ulcers, but fortunately for the battalion at this time it was
resting after action and the MO had nothing very serious to
bother about, medically speaking.

One day a rough, tough cane-cutter came to him for
attention to his badly blistered heels, acquired through not
wearing socks in his boots. The Scot eyed them and said, as
usual: 'Take two aspirin, mon.'

'Two aspirin?' the Digger repeated, looking the doctor in
the eye. 'What am I supposed to do with them? Put them
in me bloody boots?'

Humour can be shown in more ways than mere speech.
The late Rolly Goddard, when he was proprietor of the
Anzac Hotel, Amiens, liked to recall Digger humour whenever I stayed at his hotel. As a corporal of the 16th Battalion
he was detailed to escort a private to an Australian hospital
in Egypt. The Digger was thought to have a hernia and
Goddard was under orders to wait and get the doctor's report.
The suffering Digger was taken into a ward and told to get
into bed. Then two sisters arrived and placed a screen around
the bed while they examined the new patient. Goddard,
waiting at the end of the ward, suddenly heard the girls burst
into peels of laughter.

'Callous bitches,' he thought.

When the nurses came towards him, still laughing, he said:
'What's the trouble with him? Is he swinging the lead?'

'No, he has a hernia all right,' one of the sisters said, and
off they went, highly amused, to return in a few moments
with a doctor, who also disappeared behind the screen. Then
he, too, laughed loudly and long.

Corporal Goddard could stand it no longer. 'What's so
funny?' he asked the doctor as he left the patient.

'Come and look, corporal,' the doctor said and took Goddard behind the screen. The Digger lay on his stomach on
the bed, the back of his body exposed. On each of his buttocks

was tattooed a very large, realistic eye and above the eyes were the words, 'I saw you first.'

Back in Boer War days the Australian soldier made light of wounds. A Queenslander shot in the ribs, chest and throat waited patiently for a doctor to reach him as he lay in hospital. When he came to the soldier's bed the doctor asked: 'Does it hurt, trooper?'

The Queenslander croaked his reply. 'Hell no, only when I laugh.'

This joke has been told many times and is supposed to have originated with an English retired colonel who, while hunting lions in Africa, was attacked by natives and speared several times before finally being impaled to a tree. Telling the story in his club, the colonel is said to have made the reply quoted above when questioned by a fellow club member. But the story with the colonel as the central figure was not current before 1935, while the Queenslander's reply is quoted in F. Cotton's now obscure book, *Transvaal Memories*, published in 1907.

Never awed by rank, the Digger has devastated senior officers with his ironic nonchalance. When I was a sergeant in 1940 I several times drew command of the brigade guard, which meant that I was responsible, with my guard of two corporals and eighteen men, for the custody of defaulters lodged in the brigade guardhouse. All field sergeants were liable to this duty, though sometimes they were second-in-command to a lieutenant.

The tour of duty was twenty-four hours—from about 4.30 p.m. one day to the same time next day. It was a sleepless twenty-four hours for the sergeant, who not only had to ensure that his prisoners did not escape but had to feed them and keep his guard on the job. The brigade field officer could be expected to call at least once in the twenty-four hours to make an inspection and all too often the brigadier himself was likely to appear. For field officer or brigadier, the guard had to be turned out—that is, paraded in front of the guardhouse, to present arms to the officer and to be inspected by him. At this time the brigadier concerned was officious and mad with zeal. He liked to call when least expected to

make sure that the guard—and particularly the sergeant—
was on the job. Many of the sergeants who drew this duty
were potential officers and the brigadier wanted to see that
they were worthy of their prospective rank.

With trained troops guard duty was not particularly onerous, but when a sergeant had to take rookies for the job it
was one long headache. The men didn't see the sense of it,
they were not disciplined, they didn't like walking a beat for
two hours at a time on a cold night and they could not be
relied upon to challenge intruders. And even the brigadier
was an intruder until identified, and he expected to be challenged by the sentry he encountered.

One night I had a guard composed entirely of men who
had been in the army no longer than three weeks; even the
corporals were only newly promoted. I impressed upon all
sentries the importance of challenging everybody who came
near them and told them that if the field officer or brigadier
arrived they were to shout 'Sergeant of the guard!' or 'Turn
out the guard!' But I had no confidence in these boys and
most of the time I roamed the posts giving the sentries
practice in challenging me. 'The brig is bound to come some
time,' I said, 'and I don't want to miss him.'

But I couldn't be around the posts all the time and the
brigadier arrived when I was not on the scene. He happened
to approach the guardhouse across the path of a new Digger
whom I now remember only as Tich, a short, casual man on
whom even a tailor-made uniform would look like a sack.
Tich had become uncomfortable in his uniform and after my
last visit to him at one in the morning, he had undone his
collar, boots and belt, and put his chinstrap behind his head.
Also, he had decided to have a fag—a heinous crime on duty.

Still, when he saw a man in a peaked cap approaching him
in the dark, Tich remembered to say 'Halt!' (It should be:
'Halt! Who goes there?') and to come on guard with his rifle
and bayonet. Then he forgot what came next. He took a
couple of puffs at his cigarette while thinking.

'Well?' the angry brigadier said impatiently.

'Come up real slow and let's have a dekko at you,' Tich
said. (After the visitor has identified himself, the sentry should

say: 'Advance and be recognised.')

The brig came up real slow and Tich looked him over, not recognising him or his badges of rank. After only three weeks in the army he had never seen a brigadier and had forgotten his few lessons on rank identification.

'You look all right to me, mate,' Tich said at last. 'What do you want?'

The brig, apoplexy pending, said: 'Don't you know who I am?'

'Buggered if I do,' Tich confessed. He eyed the red band around the brig's cap. 'I reckon you could be in the band.'

'I'm the brigadier!' the outraged camp commander said.

'Are you, be jeez!' Tich said. 'Well, just you wait till the sergeant of the guard catches you! He's been waiting for you all bloody night!'

Organised protests are not allowed in any army, even in the most democratic, but the Diggers were always able to get around this by giving a protest a humorous air. Food or its manner of preparation has not infrequently been a subject of protest, but mostly grievances were remedied by direct complaint to the battalion or unit orderly officer, whose duty it was to visit the various mess huts at meal-times with the orderly sergeant who asked of the dining troops 'Any complaints?' A Digger with a complaint was then entitled to air it; the orderly officer investigated it and whenever justified the matter was rectified.

In one camp the men objected to the fish, which, they said, were not fresh. Three complaints were made in the regular fashion, but the fish were still on the nose. Then the troops acted. One day officers lunching in their mess were startled to hear the battalion band playing the 'Dead March'. They looked out the mess windows to see about two hundred men, in column, slow-marching around the mess, led by privates of the unit band, behind whom came four men holding, most reverently, a stretcher draped with the Union Jack.

As the astonished CO focused his eyes on the incredible scene the parade stopped and the band ceased playing. The four stretcher-bearers dug a small hole, only a few paces from the officers' mess, carefully took the flag from the stretcher and, picking up the plateful of fish that lay under it, they interred it while every man present held his nose. Nobody laughed or spoke and finally a bugler sounded the 'Last Post'.

The next fish served up were so fresh they were still breathing.

More than once a Digger has brought humour into play to cloak fear or pain or sorrow. On the Kokoda Trail in 1942 I sat by a member of my platoon who had been shot through the chest. He was in a bad way and I knew from the medical orderly's manner that the Digger had only a slim chance of surviving. The orderly had cut off the wounded man's shirt to dress his wound and I could see his chest heaving with the effort of breathing.

The Digger was about twenty-two and he was scared, but he was trying to appear nonchalant. 'This'll be a homer for sure,' he said several times, referring to his wound. He laughed. 'Those yellow bastards can't shoot for nuts.'

'Don't talk so much,' the medical orderly said.

'Gawd!' the Digger said in mock disgust. 'Even when a man cops it somebody wants to order him about.'

But he lay there quietly while we waited for stretcher-bearers. He had lost a lot of blood, the wound was near his heart and even in a city hospital with the best of medical equipment it wasn't likely that doctors could save him. I think, towards the end, he knew it. Anyway, the fear left his eyes.

He gave another croaking laugh and nodded towards the ripped and bloody shirt with the bullet hole near the pocket, lying where the orderly had dropped it. 'Blimey,' he said, 'Lofty's going to be as wild as a cut cat—that's his shirt.'

And then he died.

He had demonstrated another Digger tradition—not to 'squeal' when hurt.

During Alamein a Digger of a New South Wales infantry battalion, seriously wounded by a burst from a German machine-gun, lived long enough to reach an advanced dressing-station, where a padre found him. A padre seems to know before a doctor when a man is dying and this padre knew that the Digger's chance of surviving his frightful wounds was nil. Nevertheless, he was cheerful and he smiled as he said to the soldier: 'Anything I can do for you, Digger? I suppose you won't feel like writing letters for a while; have you a message for anybody?'

The Digger was in too much pain to grin, but he said:'Don't give me the raw prawn, padre. I'm cashing in. But I've got a message all right—tell Pongo Roberts to stay away from my sheila in Cairo.'

The last words of a Digger named Charlton, seriously wounded in a bayonet fight with a German at Bullecourt, are somehow especially impressive. By the time he reached the hands of a doctor, Charlton had lost so much blood the doctor thought he was too weak even to speak. But as the doctor was about to administer an anaesthetic, the Digger whispered, 'Doc!'

'Yes, what is it?' the doctor asked.

The Digger made an effort to speak more loudly. He said: 'That bastard could fight!'

He died under the anaesthetic.

Some Digger humour is highlighted by swearing though I knew many who never as much as said 'Damn!'. Many people have criticised soldiers for their language, but swearing is a safety valve, a natural means of expression, a way of damning boredom and of scaring away fear.

I was once in a camp commanded by a puritan colonel who issued strict orders forbidding swearing. He might as well have tried to forbid breathing, but since officers and NCOs were under strict orders to charge any man caught swearing a state of considerable tension built up, culminating in that camp's morale dropping to the rocks. Yet oddly enough the men did not resent the order not to swear as much as they missed the *relief* of swearing.

Sapper B. Dean of the second AIF, writing to his mother in 1942, described an enterprising angle to swearing.

We've got our mess hut finished now. Gee! we had some fun building it without nails. We've got a new rule too. No swearing during meal-hours. If you swear you pay a penny per word; the money goes to buy extra butter or sauce or any extras we might need. We counted up today— in two days 17s 8d. The idea isn't to stop the boys from swearing, but to get funds for the section. One chap waltzed in, read the rules and put 11s 1d in the tin; that entitles him to swear 133 times.

One of the best army padres I ever met—Padre Dransfield, a Presbyterian—was a swearer. He never swore during a church parade or whenever he was doing anything official, but in conversation with the troops he chose his language to suit his company. He was the most beloved and respected padre I knew. The troops lost no respect for him because he swore; on the contrary, they could see that he was human, that he spoke their language. When he died after the war, hundreds of men went to his funeral and one, in my hearing, conferred on him a muttered epitaph—'He was a good poor bastard.' The padre, could he have heard, would have been well pleased with this, a supreme Digger compliment.

Dransfield, who was not a young man, served in Tobruk. One day he was at a water point some distance from a slit trench when Stuka bombers came over the fortress. This was the padre's first experience of bombing and he wasn't happy about it. He started running for a trench, slipped and fell several times and finally, panting, he could run no farther. He looked up at the Stukas and shook his fist at them. 'All right you bastards,' he said, 'drop your bloody bombs!'

NCOs are supposed to use more bad language than any-body else in the army, but the fact is the inspired and competent NCO seldom uses bad language; he merely chooses his words deliberately. A bishop once wrote to an Australian training brigade commander, mentioned that he had received complaints (he didn't say from whom) about abusive NCOs and made this pronouncement: 'The recruit is entitled to civility. Commands are more readily obeyed when issued in good simple English.'

What could be more simple than, 'Get into step, block-head!' or, 'Haul that butt of yours into step, birdbrain, before I boot it clear up past your shoulder-blades!'

It will be readily perceived that this is a physical impossibility, but the recruit thus addressed, conjuring up a mental picture of himself sitting on top of his pelvis, generally did haul his butt into step. And perhaps later, in combat, he remembered that peculiar moment and made a better member of a fighting patrol.

One of the best CSMs of the AIF was Ben Brock, a tall, tough Queenslander. During rifle inspection, Brock would peer down the weapon barrel of some hapless rookie and say: 'Congratulations soldier, I'm happy to see that the hole has quite healed up now. Fall out! You've got three minutes to operate on that rifle before I operate on you.'

Except while giving commands, Brock never shouted. His comments were all the more scathing for being delivered quietly. 'That's not your girlfriend you're holding, soldier!' he would say. 'It's your rifle. Hold it, don't pet it!' Or, 'You're marching like you're in trouble, Private Brown. If you want to go to the latrine, fall out!'

Digger NCOs have long believed that official punishments for minor breaches of discipline are too soft; also they are apt to make a man resentful and give the impression that the NCO who is forced to parade a man to his company commander is unable to handle trouble himself. Consequently, the keen NCO thinks up his own punishments, always more effective than those approved by the army. I knew one who used to set defaulters cutting the grass around the camp— cutting it with fingernail clippers, one blade at a time. A Victorian sergeant, Jim Saville, had a large bottle of dried peas. Defaulters had to count them and come up with the right answer. Since there were several thousand peas in the bottle it was easy for Saville to take out or add a hundred so that one reluctant mathematician couldn't pass the correct answer to another.

Push-ups have always been a steady stand-by of the NCO. This is an exercise in which the defaulter is made to lie on his chest and then raise himself on his arms and toes, keeping the back straight. Some sergeants had a classified list of push-ups: six for inattention, ten for speaking, twenty for smoking. One NCO found that even push-ups could not keep recruits awake during a lengthy, technical lecture on the theory of small-arms fire. One day he figured out the perfect solution. Resignedly, he told his listeners: 'Okay, maybe this lecture *is* supposed to save your lives, but I'll admit it's dull. All those who feel like it can go to sleep. I'll wait just a few minutes so as not to disturb you.'

The troops stared at him. A few began to nod off immediately. One suspicious character said: 'We won't cop any push-ups?'

'I won't punish any man for falling asleep,' the sergeant said gently. 'But when a soldier does sleep the man on his left and right will come out here quietly and do push-ups until the clot wakes up by himself.'

The scheme worked too effectively. Every man was so busy making sure his neighbour didn't sleep that nobody had time to concentrate on the lecture.

There have always been those NCOs who are direct descendants of the Roman overseers who used to flog galley slaves to greater effort. One of the best of this type was WO1 Jock Wilson, one-time Regimental Sergeant-Major of the Royal Military College, Duntroon. A Scot, Jock had spent his life in the army and could instil more cold fear into a man than any NCO I ever met. I knew his powers at first-hand because I was a member of an AIF officers' training course at Duntroon in 1941. As it happened, I was put in charge of No. 1 Section, which meant that I was the marker; consequently on the RSM's command 'Marker!' I was the first on to the hallowed parade-ground. When I was in position the other trainees fell in on my left. Alone and unprotected, I came under Jock Wilson's eye at every parade.

While instructing classes of students on the finer points of giving words of command for parade-ground drill, Jock would approach a student who had just shouted a command and ask, in a friendly voice: 'And what's your name, mon?'

Coming smartly to attention, the student would reply: 'Brown, sir.'

Jock, his rat-trap mouth close to the student, would muse: 'Brown ... a verree fine name, a verree fine name'—and then, in a shattering roar—'but you've got a bloody awful word of command!'

He could make military college students—all experienced NCOs—feel as if a three-inch mortar bomb had exploded in their midst.

The most abusive sergeant I ever met was a West Australian. On one memorable occasion he started a reinforcement platoon marching under the hot sun, sat under a tree and calmly watched the platoon march until it was out of sight. Two hours later, dusty, sweating and puzzled, the men came back. The private who had assumed command halted them before the still-seated sergeant. Now he got to his feet, looked them over and let fly: 'You bunch of idiot numskulls! You pack of drongoes! Soldiers! Bloody fools! Blasted nitwits! Suppose we were in action and I was killed, what would you dopes do? Pack up and go home? You'd all be killed, that's what. When there's nobody to give you orders, give 'em to yourself.'

He eyed the man who had assumed command. 'I was going to get you two stripes, but I'm not now. You took too long to come to your senses.'

When he was promoted to RSM this operator was noted for his toughness. If a soldier collapsed in the heat on the parade-ground he would snap: 'Let him lie! Keep your eyes up! Any soldier looking down will wind up in the guard-house.' He produced good soldiers.

Not a few Digger NCOs and even officers have been known to take off all insignia of rank and invite a defaulting trainee to scrap in some concealed spot. This is strictly against regulations, but it usually results in a troublesome recruit gaining a healthy respect for the NCO or officer.

Most sergeants and warrant officers find their tongues adequate without resorting to fists. I remember a sergeant who had some slight bother with a trainee named Montague. A snob and mother's boy, Montague winced every time the sergeant called him 'Montaig'.

One day, during roll-call, Montague spoke up irritably. 'My name is pronounced Mont–a–gue, sergeant,' he said.

'Is that so, Mont–a–gue?' the sergeant said. 'Well, fall out here. I've got some fat–i–gue work for you.'

One of the main traits of the Digger character is his calmness, his refusal to get excited. 'Don't panic, soldier,' the instructor sergeant said to the rookie who in his anxiety had dropped a live grenade in the bomb-throwing bay, 'we've got a good five seconds before it goes off.'

And there was the example of Captain P. H. Cherry, VC, MC, of the 26th Battalion, when his company met strong resistance at Lagnicourt. His subsequent actions were contained in three laconic messages sent to his CO.

Message 1: Held up by strong-point. Have you any Stokes? [mortars].
Message 2 [Half an hour later]: Can't wait for Stokes: having a go at it: will report result later.
Message 3: Got them with Lewis-guns and rifle bombs from the flanks. The lot killed. Damned good.

Nevertheless, under certain circumstances, the Digger will do his block. In *Blokes I Knew* P. C. Neasbey relates an incident in Syria in 1941 concerning a Digger named Curly— the AIF had thousands of men nicknamed Curly. An officer told Curly that he was being entrusted with a lone mission and handed him a box.

'I'm going to give you a job where you will have to keep quiet,' the officer said. 'There are sticky-bombs in that case and you will take them with you back to where you mined the road and take up a position overlooking the area. If Vichy tanks appear give them the works.'

'Very good, sir,' was all Curly could trust himself to say as he turned away and tramped savagely up past the front-line. Careful lest he bring enemy bullets buzzing like angry hornets about him, he found a comfortable position behind a friendly rock high above the road and composed himself to his monotonous vigil. Dawn came, but there was no sign of the expected French attack, and as Curly opened his emergency rations and ate his breakfast he began to wonder what he was now supposed to do. Was he supposed to sit here like a mug all day waiting for relief or was he expected to report back?

An indiscreet movement and a bullet from a French

sniper as it ploughed into the earth beside him decided Curly—he was staying here until darkness came again. As the hours slowly passed he was frequently attacked by mosquitoes which he could endure, but each time he tried to ease the pain in his cramped limbs, a *ping* warned him that the sniper was still after him, and Curly's temper— never very placid—had now grown to an almost over- powering frenzy.

Then as the long evening shadows came creeping across the mountains he heard the unmistakable clatter of the tanks approaching and gleefully he opened his box of sticky-bombs—he'd teach those French cows to snipe at him! About 200 yards to the left of his position the road turned sharply behind a bluff, and Curly never took his eyes from that corner as he waited, grinning with antici- pation, for the enemy to appear. Then as the leading tank lumbered around the bend he carefully selected a bomb and prepared to give it the works.

The second tank had rounded the bend and then there came from the mountain-side above Curly a hail of small- arms fire directed against the tanks which were still a hundred yards from the youngster. Curly exploded with wrath as he saw the tanks quickly turn and head back along the road out of his range. Disregarding the enemy sniper, he bounded to his feet and shook his fist at the hidden riflemen.

'You bloody fools!' he roared. 'A man sits here like a bloody dill all day and you mugs come along and fuck the show!'

From the beginning of his history the Australian soldier has had a reputation for being quick with a retort or an answer. In 1915, when the 4th Brigade, under Monash, was camped at Heliopolis, the CO had a kit and equipment parade. Stopping when he reached Private Ginger Reynolds, he ordered, 'Show me your identification disc.'

'Ain't got one,' said Ginger.

'Do you know that is a serious crime in the army?' Monash said sternly. 'You do know, I suppose, what your identifi- cation disc is for?'

'Of course,' said Ginger. 'When I get to the front and I'm

stiff enough to get my bloody head blown off they come along, pick up the pieces, look at my identification disc and stop me bloomin' pay.'

A padre (it is said) addressed a detachment of Diggers going to England on leave. 'My comrades,' he said, 'remember that Hell is a frightful place and it's filled with drink and loose women.'

A Digger from the ranks called out piously, 'Oh Death, where is thy sting?'

Diggers are as nonchalant among the great as among their mates and perhaps because of this they manage to be the focal point of any group. Considering the drabness of the Australian uniform until relatively recent times, this is quite a feat.

One day early in the Malayan campaign in 1941 the Commander-in-Chief Singapore, Sir Richard Brooke-Popham, was standing on the wharf surrounded by many senior officers, all wearing several rows of ribbons. Brooke-Popham had the most inspiring collection of all.

A Digger who had just arrived on a troopship ambled down the gangplank and from a distance stood studying the array of fruit salad on the C-in-C's uniform. Then he started towards Brooke-Popham. Instantly everybody looked at the Digger who strolled up to the general and said casually: 'Say, mate, what rank did you get for all those ribbons?'

A little startled, Brooke-Popham said: 'I'm the Commander-in-Chief here.'

'Yeah?' the Digger said, interested. 'Well, I'm one of your troops and I'm glad to meet you.'

He held out his hand to the C-in-C who, a little dazedly, took it. The Digger nodded to the assembled brass and went back to the ship.

Diggers are not above taking a rise out of an officer if they consider he has been unjust to them. In a letter home during World War II an infantryman wrote:

A couple of boys had been on the spree. Our lieutenant found two bottles of wine in the tent and immediately put them on mess duty for a week. The first night, when setting the officers' tables, they gave the lieutenant a huge toasting fork, an ancient rusty carving knife and the largest spoon in the camp—one used for stirring porridge. He was as wild as a bull and immediately gave them an extra week fatigue. The next night the boys retaliated by giving him a baby's eating-set—tiny fork, spoon and pusher.

For all his toughness the Digger is soft-hearted and sentimental—though he would deny with his last breath that he was sentimental. Padre Fred Spence, a Presbyterian, told this story of Digger soft-heartedness.

During a heavy shower in a town in Palestine, a Jewish woman with a baby in her arms and another child clinging to her skirts had taken shelter in a shop doorway. Two Diggers, celebrating Christmas, also sheltered in the doorway of the shop, which happened to sell prams.

The first Digger, without much ceremony, took the child from the mother's arms and walked into the shop, followed by the anxious mother. To her embarrassment the Digger ordered one of the best prams and a waterproof cover, paid about £4 for it, put the baby into it and walked out of the shop pushing the pram and calling on the mother to follow.

The baby didn't mind a bit, but mother was still doubtful. Down the road they went, and when they crossed the road, the second Digger held up traffic for the pram to cross. Safely on the other side they handed pram and baby back to the mother and went their way.

Diggers have always thought highly of chaplains and Salvation Army padres were the most popular of all. Generally speaking, they were the closest to the front-line, they worked hardest and their compassion knew no bounds.

An outstanding 'Sally' officer was Adjutant Bramwell Gibbs who served in the Western Desert, throughout the siege of Tobruk and in Syria. His gallantry in the battle of El Alamein won him the British Empire Medal, and his cry, 'Here's your coffee, come and get it, or I'll give it to Rommel,' became a byword in the 9th Division. After his return to Australia he was appointed deputy commissioner for Red Shield work

with the 2nd Australian Corps HQ. Returning to Australia
from New Guinea in December 1943, he was killed in a
plane crash.

He gave something of a modest description of his work in
a letter to friends in 1942.

> Yesterday I was unfortunate enough to break the front
> spring of my bus, the 'Murwillumbah'. Today I am writing
> in the vast open spaces with nothing but sand and stone
> as far as the eye can see, with the pad resting on a biscuit-
> tin. The afternoon sun is blazing on my back. The pri-
> muses are roaring beside me under two 20-gallon [90-litre]
> tins of coffee which we are preparing for the boys. The
> wind is very boisterous and the place is vibrating with the
> roar of heavy guns. My batman has been serving the
> wounded at the main dressing-station as they come through,
> while I have been serving at the dressing-station in the
> front-line.
>
> I put the coffee in an ambulance when going up empty,
> serve it to the men and come back with a load of wounded
> and fill up and return again. The corporals I have attached
> to me are doing a splendid job and are a great help.
>
> I am now running three mobile units here in the desert.
> Last month—August—we served more than two thousand
> gallons [9000 litres] of coffee, besides biscuits and other
> comforts. It is great satisfaction to feel that one is doing
> so much that is appreciated. I feel the greater the sacrifice
> the more worth-while the effort and continue to get
> boundless joy in serving. I don't want to return to Australia
> till it is all over, but hope it won't be long. I wouldn't
> leave my boys for worlds.
>
> The other morning I pulled into a little war cemetery
> where they are burying the dead, and one of the battalion
> boys ran to greet me with 'Gee, Mr Gibbs, we are pleased
> to see you! It is reported that you were killed yesterday,
> and we thought there would be no more coffee'. I wasn't,
> and I do thank God for sparing mercies.

Writing home in 1942, Private M. J. Connor of the Army
Medical Corps said:

> One is lucky to be able to scrounge a couple of sheets of

writing-paper a week unless he comes across a padre.
Which reminds me of a Salvation Army padre here. One
night going forward with an ambulance we were hailed
by an old chap wanting a lift to the front to be among
his boys. He carried a large suitcase which we thought
contained his own gear, but later we learned that he had
left all his personal gear behind and the case was filled
with cigarettes for the troops. The last I saw of this
wonderful man that night he was holding the hands of
the wounded and giving all the cheer he could. All about
him the earth and the sky were one great big blitz, but
this padre was not afraid.

I cannot be sure, but I think Private Connor had met
Adjutant Gibbs.
Another beloved padre was Padre Frank Hartley ('Happy
Hartley'), a Methodist minister from Victoria. An infantry
captain in New Guinea told this story about Hartley.

Our Christmas mail came up through to the front-line. A
padre brought it up. He is mad, mad—but he deserves the
VC. The regiment was far forward, lying in holes, sniped
at incessantly, wet, hungry and more miserable than ever
before in their lives. Up the trail came Chaplain Hartley,
with a bag of mail over his shoulders, as unconcerned as
if taking a Sunday stroll.
    They pulled him down and asked what he was thinking
of to take such risks. His reply was, 'I knew you'd want
your Christmas mail.'
    We reminded him that there were snipers. With a
matter-of-factness that amazed us he said: 'Now that you
come to mention it I think I was sniped at two or three
times on the way up.'
    'And did you lie down then?' we asked.
    'No, I thought if I was to be killed I had rather be
walking along than lying with my face in the mud.'
    He stayed with us, attended to the wounded, brought
in many wounded men and on one occasion when our
men weren't game to move a finger he went out to a man
we knew was hit and most probably killed. He found his
body, dug a grave and buried him, standing cap in hand

to read the service—and, as we found out later—a Jap machine-gun not thirty yards away.

The regiment's love and admiration for a brave man found expression when every man who was able, regardless of faith and denomination, attended the last church parade he held before he left New Guinea.

The Digger is supposed to be girl-hungry and he probably is; soldiers away from home tend to lack emotional restraint in the face of loneliness and sexual drive. The Diggers of World War I, World War II, Korea and Vietnam have been no more predatory than the Australian soldier of the Boer War, who showed a keen interest in the girls of whatever town they happened to be in. At least 200 troopers of Australian contingents were discharged in South Africa to marry South African girls.

The diary of the scion of a well-known Queensland pastoral family contains frequent references to the trooper's amorous excursions. On one occasion he wrote: 'It was embarrasing [sic] today [12 September 1901] when three girls who were with English soldiers left them to join Bruce and Howard and me. For some reason the women always seem to prefer Australians, though I must say the Canadians waste no time!'

The Diggers in France wasted no time either. E. J. Rule describes the scene in a rowdy estaminet:

Two rather nice-looking French girls were handing around drinks to a crowd of Aussies. The tables were close together and one of the girls, in broken English, kept calling out: 'Deeger, do not touch ze legs with ze hand,' and occasionally making leaps and bounds like a young kangaroo. By the time I left the girls were behind the counter and refused to come out.

The Diggers of World War II were no different—they had an eye for a pretty leg or for a rounded curve and if possible they wanted to touch what they saw. But not all Diggers had roving hands, even if all did have roving eyes. A vast number of them, while interested in the women with whom they came into contact—Egyptians, Lebanese, Jews, Eurasians, Greeks, Chinese, Malays and so on—never once gave way to the desires which haunted most of them, though their abstinence had a number of reasons.

'It's this way,' I heard a Digger say. 'I wouldn't mind going to bed with a sheila, but I never did like to hurry or share my meals.'

But opinions differ. There was another Digger who said, as he eyed a flirtatious Jewish girl in Tel Aviv: 'I wouldn't touch her with a barge pole! Why should I: I've got two good hands, haven't I?'

The Digger officer retained the sense of humour he had as a private, though his senior rank gave him less opportunity to exercise it. Still, this was not the case with an infantry captain who, after a lot of front-line service during World War II, was sent to an infantry training battalion in Australia. He had been wounded and also had suffered a severe illness and was no longer considered fit for active service. As a company commander with the training battalion he was bored and uninterested, not unnaturally considering his background, although experienced officers and NCOs were of great importance in the vital work of training reinforcements.

This particular officer, however, was never cut out to serve anywhere except at the front; he was a man of great energy and dash and he loved to pit himself against danger. There was very little danger in a training battalion and the paper warfare in which he now had to indulge irritated him. It seemed to him that he spent most of his time signing useless forms.

One day, bored to the limit, he decided, for his own amusement, to introduce a new form. A large imposing document when drawn up, it was headed 'DAILY SUMMARY OF FLIES KILLED IN COMPANY MESS HUTS'. There followed various sub-sections such as 'Flies Killed by Sticky Paper', 'Flies Killed by Swat', 'Flies Killed by Spray'—and so on. Columns gave not only the number of flies killed by the various methods, but also size and type of fly—small, medium, large or very large; ordinary, blue-arsed, green-backed or blow. The term 'blue-arsed fly' had significance in army language, for it was often said of an officer or NCO that 'he buzzed around like a blue-arsed fly'.

Impressed by his creation, the captain gave the form a number, in army style, and had a number of copies duplicated. Each day after that he filled in a form and sent it off direct to Brigade HQ, listing his company and battalion and signing the form in the usual manner. Occasionally he added under the heading 'Special Remarks' such intelligence as: 'It was particularly noticed today that flies of the green-backed variety are drawn towards plum jam. I suggest that the Quartermaster-General be informed of this with a view to making the said jam less appetising to flies.'

About three weeks after the captain had begun his regular reports two of his fellow company commanders called at his orderly room. One of them said: 'Jack, have you been in trouble with Brigade HQ over Fly Reports?'

'The daily summary of flies in mess halls?' the captain asked seriously. 'No, I've been sending them in every day. Do you mean to say that you've been giving them the go-by?'

'We've never heard of the damned report,' the other officer said. 'And my orderly room staff haven't even seen it.'

'Well, don't panic,' the captain said. 'I've got some spares you can have.' He supplied his brother company commanders with report blanks for a few weeks until wads of blanks began to arrive from Brigade HQ. He used to amuse himself by wandering into mess huts around the camp to watch conscientious orderly sergeants and orderly corporals counting dead flies. Base records improved on the captain's original form, but then, as he said, the armchair warriors at Records were experts in time-wasting; he himself was only a novice. Records included a section in which company commanders had to estimate what percentage of the total number of flies in each mess hut had been killed.

The captain never did find out how widely the form spread through the army, but he was sorry later that he had not named the report after himself as a bid for immortality.

Inspired with success, he started a new form called 'WEEKLY SUMMARY OF TROOPS' ACTIVITIES ON LEAVE', listing, by percentages, soldiers' preferences for female company, beer, gambling and 'intellectual stimulation', among other things. It was a startlingly frank and enlightening summary, but a few weeks after initiating it the captain was discharged.

Women ... emotions ... rations ... death ... boredom—the standard Digger rule was to treat everything the same way; that is, swear at it or laugh at it, or both.

26

'Extraordinary Fellows'

This book, in explaining the evolution of the Digger, is not an attempt to glorify war. But as wars have occurred and Australians have taken part in them it is only right that their service, suffering and sacrifice should be recorded. If this has resulted in their courage being glorified, so be it.

New to the grim business of war when most of the countries of the world had been in periodic conflict for centuries, the earlier Diggers proved themselves not only equal to but better than the troops of the nations with military traditions. Their successors in later wars, up to and including Vietnam, showed this same prowess.

To a greater degree than any other troops of any period the Diggers have shown what comradeship really is, though they call it mateship. In the contingents which went to the Boer War, in the AIF of both world wars and later in the RAR mateship had the force of a religious creed, whether in action or not. The camaraderie of Australian troops was one of their unique qualities, one which contributed much to getting Diggers known and talked about.

Another outstanding Australian Army quality has been its leadership, its revolutionary system in 1914–18 of promoting men to command simply because other men would follow them confidently in battle. Promotions were made regardless of the prospective officer's social and educational standing. Influence might at times have resulted in commissions but influence never *held* posts for men commissioned in this way, if they could not hold the confidence of the men. The Australian Army has always been as democratic as any army can be. Moulded on strict disciplinary lines—for no army can function efficiently without discipline—the Australian Army is built on mutual respect between officers and men and tempered with tolerance and humanity.

Not all Australians who served with the various war-time armies were good fighting soldiers for many men are by temperament unsuitable for service in the infantry and other arms of an army's cutting edge. Being in supporting units did not put them out of danger; during a walk through any Commonwealth War Graves Commission cemetery a visitor will see the graves of men who served in the medical corps, in transport, service and supply units.

During World War I the Diggers were tested more severely than most of those who experienced active service in the Boer War, World War II, the Korean War and the Vietnam War; the Diggers of the later wars faced a greater variety of hazards and types of weapons. The Owen Stanleys campaign was as arduous as any in Australian experience but we who were there did not have to endure days, weeks, months and years of heavy shelling; we did not have to charge repeatedly across open ground in the face of heavy machine-gun fire.

There never was a war like the Great War. There was much more fighting, as distinct from manoeuvring, in the Great War than in any other war in history. It was mostly a series of terrific battles fought over relatively small areas of ground and the soldiers of both sides stood almost toe to toe and slugged it out. For weeks on end they had to stay put in mud and slime under unremitting heavy artillery bombardments and gas attacks. An infantryman's chance of surviving the war was so slight that most front-line troops had accepted death long before it came; dying, some of them expressed surprise that they had lasted so long.

Others, writing home to their closest relatives, doubted if they could hold out much longer. Some, under tremendous stress, just wanted to be 'knocked out' and have done with a life which was really a living death.

The Boer War had been a war of movement. It was, in fact, even more fluid than World War II and the Diggers involved covered greater areas of country than those who served in the Western Desert and Syria in 1940–43. The Boer War Diggers made their war an individual one, where a good soldier had every chance of bringing himself through, simply by pitting his wits and his skills against the enemy. Provided he wasn't committed to a British-planned frontal attack, he was a good insurance risk. The living, with its long marches, was as hard as the fighting, but the Diggers of those days were some of the toughest men Australia has ever produced.

World War II, a war of movement, freed the Diggers from trenches and prolonged artillery bombardment, but the new and dangerous hazards of dive-bombers and strafing fighter planes, fast-moving tanks and motorised infantry, self-propelled guns, flame-throwers and rockets, vast minefields and jungle warfare, forced them to develop new and different skills from those used by their fathers two decades before. And develop them they did, with the rapidity and completeness that is yet another Digger characteristic. Later, in Korea, these skills made the Diggers the most colourful and dramatic troops of all those involved against the Chinese and North Koreans.

In unspectacular, unpublicised ways the Digger of the Emergency and the Confrontation proved himself reliable and professional. And in Vietnam the Diggers showed their American allies of the world's most powerful army that even an enemy as skilful and as determined as the Viet Cong could be beaten. Perhaps one day the Australian community as a whole will give these particular Diggers the respect they deserve.

Not all Australian soldiers deserved the sobriquet of Digger, which is an honourable and an earned title. The infantry, artillery, engineers, signallers, pioneers, paratroops and commandos, ambulance and first aid men, many military police on forward duty, drivers, Light Horse and mechanised cavalry—these men were Diggers. They fought and they were under fire; they attacked and when themselves attacked they stayed put.

In 1943 an order was given that any militia unit in which three-quarters of the men were volunteers would be described as a unit of the AIF. However, the non-volunteers were often not 'Digger material'. As an instructor I had much to do with training some of these men in 1944; although they wore AUSTRALIA on their shoulders they did not want to know how to become good soldiers and they had no intention of doing any fighting. Sometimes the conscript Australians seemed to be almost a different race of men from the volunteer AIF men.

Just what *is* a Digger, individually? Is there a typical Digger? Perhaps not, for the Digger is just an ordinary Australian. There can be no typical Digger the way there can be a typical Tommy, because Australia has never had a professional military class. The Digger was a volunteer. A clerk, pastrycook, rouseabout or businessman one day, he was in uniform the next day and a Digger a few months later.

For this very reason—that is, his lack of military background—the Australian soldier is a phenomenon; he is something the world is not likely to see again. (A close equivalent is the Israeli citizen-soldier; Israel has a small regular army. The Israelis, an unwarlike people with no military tradition before 1948, produce soldiers equal to the best in the world.) As a civilian, the Australian is easy-going and soft-hearted, he hates militarism and regimentation and he believes in live-and-let-live and a peaceful life for everybody. His whole background and way of life, both individual and national, speaks for his belief in these things and underlines his casualness.

The Digger hates regimentation, but he will endure it. He loathes war, but he will fight like a fury. He dislikes intensely being away from his womenfolk and children and he (or many of him) will go AWL to see his family. He will attack his enemy savagely and bayonet him without compunction while he continues to fight. When the enemy surrenders the Digger will give him his last cigarette. He is physically tough, but he is sentimental and will cry over his mate's body (if nobody is looking) and play gently with the battalion puppy mascot or a couple of foreign street urchins. He doesn't want to die and has a fierce joy of living, but he is not afraid to die and will go out with a grin or a gag. A man who enjoys every aspect of the physical side of life—sport, activity, love-making—he will not whimper when he is mutilated by shell-burst or explosive bullet. At the end of his tether, physically or mentally or both, when his heart is breaking, he will show no particular sign of stress or emotion, because he would feel a bit of a dill if he did, because a man should be able to take whatever is coming to him. He is not religious and discusses religion in an embarrassed self-conscious way.

He dislikes parade-ground drill, but when challenged by an understanding and respected officer or NCO to 'put on a good show, blokes, and prove to the CO this is the best platoon in the battalion' he will drill with the precision and snap of a Grenadier Guardsman. He professes to dislike au-thority, but he will crawl a mile through wire and fire for an officer or NCO he trusts. He will often refuse formal promotion because 'I don't want to be responsible for the lives of my mates', but when his platoon commander and NCOs are knocked out in battle he will assume command instantly and lead the platoon or section or even a company as competently as his erstwhile commanders—and his mates, recognising guts and ability and command, will follow him as far as he wants to go.

He says he is not mechanical-minded, but in a few minutes he can strip an enemy weapon, vehicle or radio set he has never seen before, find out how it works and get it into action. He hates to be taken prisoner and will escape if he can, even at great risk, but if escape is impossible, he will settle down and make the most of captivity.

The Digger's greatest hate is 'being buggered about', but wars and armies being what they are he gets his share of it. This leads him to grumble, but he takes few of his own complaints seriously.

The Digger flares up quickly, sometimes over little things, but his anger just as rapidly subsides and he rarely holds a grudge. There was an occasion when a Digger named Darkey Smithers left the water near Anzac Cove after a night-time swim. He came upon another Digger in the act of drying himself, dropping a lot of water on Darkey's heap of clothes in the process.

'I say, mate,' said Darkey, 'what sort of bastard are you? You don't want to be a shithead all your life!'

'Why, what's the matter?'

'Matter? Blimey, can't you see you're wettin' all my bloody clobber?'

'I'm awfully sorry.'

'Sorry me flamin' aunt, what's the good of being sorry. You ought to be careful. For two pins I'd dong yer.'

This remark was met with discreet silence. The offender donned his pants and Darkey, having quietened down, casually observed, 'Them's flash pants you're wearin'.'

This was also received in silence but when the stranger got into his tunic Darkey caught a glimpse of a crossed sword and baton on the shoulders. With a closer look he saw that the other man was 'Birdie'—General Birdwood.

'Gee, struth mate, I'm sorry, sir,' Darkey said in confusion.

'Oh, that's all right,' Birdwood said. 'The mistake was pardonable in the dark. But there's one thing I'd ask—don't call me a shithead!'

Apart from showing something of the Digger's quick temper the incident also indicates why Birdwood was popular among the Diggers; he was prepared to speak their own language.

The Digger dislikes pomposity, as numerous incidents in various wars have illustrated. At the horse trough in the town square of Becordel on the Somme front, for instance. Late in 1916 British artillery battery horses, under the supervision of a regimental sergeant-major, were being watered with pomp, ceremony and fuss, and, to a lesser degree, water. Suddenly a dishevelled Digger arrived with two mules—with the unostentatious manner of a cyclone, according to the Digger who told me the story. He pushed his mules in among the battery horses and let them drink.

The RSM, scandalised by the Digger's unmilitary behaviour, said sternly, 'What the bleedin' hell do you think you're doing there?'

'Watering a couple of mules, Dig,' the Australian said, adding as an afterthought that the RSM must have bribed the medical officer to accept him for the army with the poor eyesight he appeared to have.

'What!' shouted the very senior NCO. 'Don't you know that you are addressing an RSM?'

'Cripes. I'm pleased ter meet yer,' said the Digger affably. 'I knew yer brother, the YMCA.'

The Digger makes only one great demand—and that is for a fair go. The officer says to his platoon: 'You give me a fair go, and I'll give you one.' And the private says: 'Just give us a fair go, sarge; that's all I want, a fair go.'

He wants a fair go at the enemy, at two-up and at opportunities for leave. A fair go means that he doesn't want to be pushed about unnecessarily, made to stand in a queue for pay or for a medical parade if it is possible to sit down, be given cold coffee after a night exercise when it is no more difficult to be served a hot drink, or rostered out of turn for fatigue duty. A fair go also means that he wants to be put in the picture, to be told what is going on when he is expected to risk life on a patrol or in a major attack.

Finally, a fair go means that the Digger expects others to be fair when confronted with logical argument. Again, the Western Front provides a perfect example. When the Germans retreated before the Allied offensive of August–October 1918 they left behind several dumps of coal. Many British units and individual soldiers were soon scrounging it in readiness for another bleak winter. One 2000-tonne dump was disappearing so rapidly that a guard was placed on it. A British colonel and a captain passing the dump saw a big Australian busily shovelling coal into a wagon. The dump sentry, with rifle and fixed bayonet, was marching up and down a few paces away.

The colonel stopped and as the sentry saluted him he said, 'What are you supposed to be doing?'

'Guarding the coal dump, sir.'

'Then what about this Australian? Has he any authority to draw coal? Did he show you a chit?'

'No, sir,' replied the sentry. 'I thought that as he had an army wagon, it would be all right.'

'Upon my Sam!' the colonel said, astonished. Then he tackled the Australian. 'What authority do you have for taking away this coal?'

The Digger rested on his shovel, pushed back his hat and wiped his brow. 'I don't need any authority,' he said. 'I bloody well fought for it.' And he went on shovelling.

The colonel began to speak and then changed his mind and walked off. 'Extraordinary fellows,' he said to the captain.

All these traits apply equally to privates and officers, even to Digger generals, because many of them were also war-time volunteers.

Some Australians say, with a self-satisfied smirk: 'The Australians might not be the best soldiers in the world, but they are the best fighters.' This is a compliment wrapped around a slur. There is more to a war than fighting. Life in the infantry has been described as long periods of monotonous boredom punctuated by short periods of intense activity, and this is largely true. Hence, a soldier needs to be more than a fighter. The Digger is as good a soldier as he is a fighter. Prove to him that parade-ground work and rifle drill greatly help to develop inner self-discipline which may easily save his life in action and he will become a good parade-ground soldier. The good soldier, who is a being quite apart from the robot-like beings of so many armies, needs a high standard of self-discipline, discipline to command, and physical and mental fitness. He must have a thorough knowledge of his weapons and the use of ground and camouflage, and be the possessor of a hundred skills.

If the Digger has a serious weakness it is impatience, but impatience is something of a national characteristic and it would be as pointless to label it a fault as to call the innate reserve of the Englishman a fault.

It is true that some Australian soldiers have disgraced their uniform though extremely few have ever been guilty of a field offence, such as desertion under fire. Some have drawn time in a detention barracks for being AWL, abusing an officer, being drunk and disorderly or some other mis-demeanour, but after all the AIF never asked its volunteers for a character reference on enlistment. It asked them only to be willing to fight and frequently the man who was often in the guardhouse when his battalion was out of the line turned out to be an exceptionally fine fighter in the line. Many a Digger's offence has, on inquiry, been found to be based on his impatience and boredom with life in a static camp.

It is one thing to sum up Digger characteristics, but what do they mean to the man who has not been a Digger, to the woman who can know little of what action is like?

27

Rain, Sweat and Tears

In the early 1950s I wrote the short story that follows, 'Rain, Sweat and Tears'. It is set during World War II and illustrates the Digger ethos.

None of the men spoke as they trudged in the mud and rain along the tunnel-like track, with pale green light filtering through the armour of trees and the mad tangles of vines.

Not a man looked about him into the jungle thickness except perhaps the scouts in front whose eyes moved restlessly about the teaming oppressiveness because they were alone and consequently jumpy.

The main body of the patrol, a full platoon, was not taking quite as much care on the way back as it had on the way out and this was only human; the men had been on the track for five hours. For two hours of that they had been fighting. Half a dozen of them were walking wounded, men with shattered arms and shoulders. But they hadn't lost a man and they'd killed some Japanese, though how many they didn't know. It never was possible to know in the jungle.

At the rear of his patrol—there because during a withdrawal it was the most dangerous spot—Captain Bartholomew Black looked along the line of dirty-green-clad men, watching their boots lift mud from the track and the rain ricochet from their tin hats. Like the others he was drenched, but was unaware of it, the exhilaration of the patrol still monopolising his senses. He liked patrolling. It gave a man a job and left him alone to do it. It was a little war within a war. 'Harass the Japs on the ridge beyond Myopi,' the order had been and he had done just that. He had gone in firmly, hit hard, got out quickly and fought off pursuit.

The rain was falling like a barrage, blanketing the trail with a sound like tearing paper as it crashed through the roof of trees and foliage. The men swore at it quietly but Black was glad of it. It was almost as good as a smoke-screen in making a getaway and was now so thick he couldn't see half-way along the line of men. But still he could smell cordite and sweat in his nostrils; nothing ever seemed to eradicate *that*.

He grinned, the lines of his face running water like miniature rivers. He was pleased with his men. This platoon made a good patrol, but some other must go next time. It wasn't fair to inflict the brunt of patrolling on any one group, but it was a temptation to him when this platoon was led by Dick Weston.

He and Dick were a good team—and good friends— though affection was something Captain Black did not permit himself. They'd been privates together at first. Then Black was made a corporal and all the way along had kept one step ahead of Weston through the desert and Greece and now Papua. Black had forgotten how many patrols they had made as commander and second-in-command.

But he knew why he was a company commander and Weston still a lieutenant. Dick was a good officer but a platoon was the limit of his leadership. Black could have commanded a battalion and knew it, but that would have ended his patrolling and all the other things a company commander could do where a battalion CO could not because of his greater responsibility.

Black knew he was hard. 'Bart the Bastard', the men called him and he was proud of the name. He was relentless. Many officers might allow a minute or two rest after a position was taken but Black flayed his company into activity.

'Consolidate and exploit,' he would tell his officers and NCOs, and though this was simply echoing the military tactical handbook Black was no book soldier. Nobody loved him, Black knew. But he knew also that all his men respected him as a soldier and trusted him. He didn't ask any more. He would have considered affection a sign of his own weakness.

His was the only hand in the patrol, apart from those of the forward scouts, which never left the butt of his Owen-gun, but even he conceded that the patrol was as good as over. They were returning by a side track that looped away from the general combat area and came into their own lines directly on the flank. Still, you never knew with the Japs, ambushes were likely anywhere.

He put out his tongue and caught some water running down his face, swilled it around his mouth and let it trickle down his throat. It was strange that the terrain of war was almost in extremes, he thought. The barren, scorching, waterless desert, the lush savage profusion of the jungle where a man lived in water.

The trail wound around the ridges, dipping into head-high patches of kunai and winding up again into the insane tangle. Black's eyes flickered over it, suspicious of a bend that would make a perfect setting for an ambush, alert mind sizing up the country for a stand. He saw a figure waiting by the track and knew it would be Dick Weston dropping back from the lead for a moment for a word with him.

A dozen paces away and he could make out the broad grin on the lieutenant's face. As Black approached him Weston jocularly threw up his hand in a mock salute. 'Well, Captain Black,' he said lightly. 'We—'.

Then the crack came, muffled by the rain, and for a second Black was staring at the gouged-out space where Weston's left eye had been. The lieutenant, swaying a little, staggered back and fell loosely into the mud.

The jungle quivered as shambling, tired men leapt off the track and into cover. Section commanders prepared for am-bush, but no other shot came. The bushes settled again. Weston's tin hat lay on its crown in the mud. Black had dropped, too, and lay beside Weston. The young lieutenant's face was curiously blank and his one eye stared upward. Blood from the obscene hole was being washed over his face by the rain. Black pushed Weston's head round. There was a hole at the base of the skull.

'You bloody fool,' he murmured. He was thinking of Weston's salute, always an invitation to a sniper because it showed that the man saluted was an officer. Japs had been known to let hundreds of men go past to make sure of eventually killing one unmistakably an officer. That bullet had been meant for him, Black had no doubt. The Jap wasn't to know Dick was an officer. Badges of rank weren't worn in the jungle.

'You bloody fool,' he said again. Without raising himself he took Weston's Owen-gun and rolled off the track into the brush, feeling the slimy mossy rottenness beneath him. Feet pounded briefly and the platoon sergeant, Pack, fell down beside him.

'Lieutenant Weston dead, sir?' he panted.

'Dead as a dodo,' Black said unemotionally. He looked at the still figure on the track, arms outflung, one leg twisted beneath the buttocks.

'Think it's an ambush?' Sergeant Pack said. 'The scouts haven't spotted anything.'

'No ambush,' Black said. 'Just a sniper and he's behind the patrol. Weston was facing that way when he was shot.'

'I've got both ends of the track covered just in case,' Pack said. He was a good sergeant, Black thought; a good man in a jam. He would make a good officer in Weston's place, but he had already twice declined a commission. 'I'm just one of the boys,' he had said.

'Get the sections to paste that part of the jungle, sergeant,' Black said indicating the area. 'Tell them to throw everything they've got at it. They won't hit the Jap but he'll keep his head down while you and I get Dick's body off the track.'

The sergeant sent a runner to the leading section commander and called out orders to the other two corporals. Competent, steady firing began—machine-gun, submachine-gun and rifle fire. The jungle shook and vines and branches, cut through, fell and dangled in other vines. Black and Sergeant Pack raced out, grabbed Weston's body and brought it in to the cover of the foliage. The firing ceased. The rain kept on.

Black said: 'Take the patrol back, sergeant. Work them alongside the track for a hundred yards or so before you strike the track again. You can't mistake the way. Take all the left-handed turns. You'll hit "C" Company's left flank in about half an hour. . . . Before you go, get me a rifle and a dozen rounds.'

'What are you going to do, captain?'

'I'll follow you. Beat it, sergeant.'

He heard the sergeant shout to the section across the track, saw men pick up Weston's body and move carefully through the jungle, away from him. A rifle nose-cap tugging at a vine could draw another shot. Pack handed him a rifle and bayonet and he gave the sergeant his own and Weston's Owen-guns.

'See you later, captain,' Pack said. 'Suppose I stay with you? The corporals could take the sections back.'

Black said mildly: 'Thanks, but the new platoon commander will need you. Don't let anybody stay behind, sergeant.'

He watched until the jungle and rain swallowed up the sergeant, then turned and considered the problem of the sniper. His hand shielding his eyes from the rain, he methodically studied the limited landscape. No movement. He stuck the bayonet in his belt and began to move slowly through the vines, stopping after each pace to peer again into the tops of the trees, for every step in the jungle meant a new view. He wondered if there might be two snipers, one to cover the other.

In ten minutes he had covered as many yards. While he moved he was unconscious of the rain but when he stopped, watching, he seemed to become almost part of it. It streamed down his skin and his boots were full of it. His armpit shielded the bolt of the rifle, which he held pointed down to keep the barrel dry.

He stepped on a log and it smashed into a soggy mess. A vine caught on his hat and he slowly untangled it. His sweat mingled with the rain. A huge crimson flower stared at him and he thought of Weston's blood. . . .

He had lost sight of the track now. And he had lost sight of the thoughts he had had for the few minutes after Weston's death. Every sense was now alive and vibrant, taut as violin cords. He was concentrated on the job, and revelling in it. For the few minutes between Dick's death and when he and the sergeant had brought his body to cover, he had thought of other things—brief images of other patrols, of the desert and the mountains and of leaves and women and the thousand and one things in the life of a soldier away from home.

But these were gone now, absorbed into his resilient brain as completely as the sodden ground absorbed the inches of rain. He felt no anger, no personal desire for vengeance. These things weakened a man and occupied senses that needed to be free and bayonet-sharp.

He pressed against a tree and a large green insect fastened to his hand. He shook his hand to flick it off.

Crack!

Unhurt, he darted his glance into the trees, but saw no movement. The Jap couldn't be far away. Even from high-up the maximum vision couldn't be much more than fifty yards. Probably less. And he knew the angle from which the bullet had come.

He dropped to his stomach and inched along with his chin, belly, loins and toes on the ground, careful to keep his rifle out of the mud.

He panted heavily. Sweat ran into his mouth and his heart pounded. His lungs were tight and strained. When he stood up again five yards farther on his hands were covered in mud. He held them out and watched the rain splash the mud away. He wondered why his stomach felt bruised and remembered that the bayonet was still in his belt. He fixed it to his rifle.

He looked up again, into the tangle. And there was the Jap. Thirty feet up a tree not ten yards away. Even then the outline of the narrow sniper's platform was vague in its camouflage. He couldn't see the Jap himself. But as he watched, his hands slowly raising the rifle, the Jap's head moved just as slowly over the edge of the platform. Black almost grinned; the Jap had lost him and was wondering where he was.

He took a long cool aim at the head, squeezed the trigger, felt the shock of the rifle against his shoulder and saw the

Jap snap back and fall out of the tree, crashing down limply to the ground.

Black had reloaded automatically. He stood for an instant before his dancing senses urged him to twist about, and the shot fired at him by a second Jap only half a dozen paces away missed him by inches. For a second that hung suspended in time they stared at each other. A second that recorded and wove together a hundred images of face, appearance and background.

Then the Jap was leaping, running in with bayonet lifting. Black fired from the hip and missed. He took the Jap's bayonet on his rifle stock, side-stepping to the left and slashing the Jap across the face with his bayonet blade. The Jap seemed not to notice. He leapt two paces and thrust at Black's exposed ribs and Black chopped down with his butt and jumped backward. He needed only a second to reload but the Jap didn't give it to him.

His hat fell off and rain stung his eyes. He took the Jap's bayonet on his stock again and forced upward, so close to the yellow man he could smell his stench and see the stunted teeth.

The Jap gave way suddenly, fell on his knee and drove the bayonet upward. Desperately Black pivoted away but felt the bayonet catch and tear at the flesh under his left arm. The Jap was grinning a little, jumping in for a point.

Black saw his chance and took it, slashing down on the Jap's knuckles and feeling the blade bite. He took the point of the other's bayonet on the stock again and following through drove his own point full into the yellow face, feeling the jar as it bit bone. The Jap screamed shrilly. The rifle dropped from his fingers. Black drove his bayonet to the Jap's throat, but the little man's clawing fingers deflected it to his own stomach.

Black felt the blade push through clothes and skin. The Jap screamed sickeningly and doubled up, his hands tearing at the bayonet, trying to pluck it out.

Black jerked at the rifle insanely and when it came away, red with blood, he felt a wild pulsing surge through him. He kicked the Jap in the face and as he staggered over, drove

the bayonet in again, and again, as the yellow man lay
writhing on the bloody ground.

Black was sobbing with effort and his hair hung down
over his forehead. The Jap was still now, sprawled gro-
tesquely, his wounds seeping blood. Black swayed a little and
vomited and then sat down bside a tree and cried, defences
worn down now and thinking of Dick Weston. The tears
and the sweat and the rain ran together and he didn't care—
for he knew nobody was watching.

After a few minutes he got up, reloaded his rifle, took the
two Japs' rifles to pieces and threw the parts away, searched
the bodies without finding anything useful and set off down
the trail.

He was thinking: God help the patrol if they hadn't
cleaned their weapons before settling down to rest. He would
rouse them out and see they did it and reprimand Sergeant
Pack for his laxity.

And he knew what he would say to them. 'You lazy
bastards, what do you think this is? A Sunday School picnic?'

But he would grin as he said it and the men, grumbling
and grinning, would set to and clean their rifles.

General Index

Index of Names